STEPHEN J. ROGOWSKI

PROBLEMS for COMPUTER SOLUTION

TEACHER EDITION

Creative Computing Press   **Morristown, New Jersey**

Library of Congress Number: 79-65705
ISBN: 0-916688-14-3

Manufactured in the United States of America
Third printing March 1980
10 9 8 7 6 5 4 3

Computer enthusiasts will also want to subscribe to **Creative
Computing** magazine, the #1 applications and software
magazine. Subscriptions in the USA cost $15 for 12 issues;
foreign $23. Sample copy $2.50 anywhere. Write to:

Creative Computing Press
P.O. Box 789-M
Morristown, NJ 07960

Contents

THE AUTHOR

Stephen John Rogowski is affiliated with the State University of New York, Albany, New York. He has previously published several articles on the use of computers in education and conducted workshops in computer usage for the National Council of Teachers of Mathematics and other groups. He has also published various plotter designs in the form of posters and T-shirts.

Preface:

This preface will serve both editions of this book. There is a student edition which contains problems to be solved, references, and an appendix of useful information.

The student edition is designed to encourage research and preliminary investigation on the part of the student. The problems are ordered by subject area, i.e., arithmetic, algebra, geometry, etc. Certain problems can be expanded, or shortened. References are given in order to encourage preliminary research.

The Teacher's Edition contains solved problems and has the following features:

(1) The student problem page is reproduced;
(2) Actual problem solutions, as printed by the computer, are given;
(3) A program which produces this solution is shown;
(4) The analysis, for most problems, is intended to make clear to the teacher exactly what went into the program, to explain any algorithm used, to give further references, and occasionally suggest further reading or research.

In some sections, more than one program has been listed. This approach is taken either to show an alternative way of solving the problem or as part of stepwise, multiple-program progression toward a solution.

The reader will find that some problems have not been either analyzed or solved. Special interest problems or problems which have never been solved, are posed to give the student an opportunity to deal with some of the unsolved problems in mathematics. Some research and an attempt to solve these will sharpen the student's insight and awareness.

The book can be used with almost any computer-oriented course of the high-school or college level. Any programming language can be used to solve the problems. However, all solutions are given here in BASIC. BASIC is the most popular and easiest to learn of the programming languages used in education.

Many problem solutions were written in EDUCOMP BASIC and run on an 8K (word) Digital Equipment Corporation PDP8 computer. Other problem solutions were implemented and run on the UNIVAC 1108 at the Computing Center of the State University of New York at Albany (SUNYA). The software was a Real-Time BASIC (RTB) package authored by personnel at the center

Some of the programming statements used are not available on all BASIC software systems. By the same token the BASIC being used by the reader may have features which were not available to the author. An attempt has been made to point out these differences within the analysis following the program in question.

This project was made possible through a grant from the SUNYA Computing Center. I wish to extend my appreciation to the staff of John Tuecke, Assistant Director for Academic Services, at SUNYA for their assistance. My thanks also to my student assistant Dave VanSchaick, who did much of the drawing and programming. Mr. Brad Longdo of the Media Center at Waterford-Halfmoon High School helped me in planning and designing many aspects of the volume.

Stephen John Rogowski

Blind reliance on automobile speedometers can sometimes cost one a ticket. They are strikingly unreliable, in some cases being off by as much a 15 m.p.h. The trooper, when confronted with inaccuracy as an excuse for speeding, says *Tell it to the judge*. The judge says *Pay the fine*.

An ideal way of checking your car's speedometer would be to time the car over a measured mile (such as those between distance markers on superhighways). If the speed is held constant, a simple table should allow one to convert the elapsed time for the mile to a speed in miles per hour.

Program the computer to make such a table. Units are important here. Make your table from 40 seconds to 70 seconds in divisions of one second. Be sure to give the speed in miles per hour.

You may want to replace the teletype paper with a ditto stencil and distribute the chart to your less law-abiding friends.

```
FAULTY SPEEDOMETER SPOTTER

SECONDS        MILES PER HOUR
*******        ***** *** ****

70             51.4286
69             52.1739
68             52.9412
67             53.7313
66             54.5455
65             55.3846
64             56.25
63             57.1429
62             58.0645
61             59.0164
60             60
59             61.0169
58             62.069
57             63.1579
56             64.2857
55             65.4545
54             66.6667
53             67.9245
52             69.2308
51             70.5882
50             72
49             73.4694
48             75
47             76.5957
46             78.2609
45             80
44             81.8182
43             83.7209
42             85.7143
41             87.8049
40             90
```

```
10 PRINT 'FAULTY SPEEDOMETER SPOTTER!!'
20 PRINT
30 PRINT 'SECONDS','MILES PER HOUR'
40 PRINT '*******','***** *** ****'
50 PRINT
60 FOR T = 70 TO 40 STEP -1
70 LET F = T/3600
80 LET R = 1/F
90 PRINT T,R
100 NEXT T
110 END
```

ANALYSIS

Line 70 converts from seconds to hours and does the computation.

Since $D = R \cdot T$ it is obvious that $R = D/T$.

The distance is 1 mile; hence, the numerator in line 80. F is the number of hours and R is the rate.

Printout is to six places and relatively useless to that degree of accuracy. It is left as an exercise to reduce the output to a more meaningful number considering present day speedometers.

3

2   Paper Folding Problem

Suppose you were to fold a piece of paper a whole bunch of times. Each successive fold should produce a piece of paper twice as thick as the previous one.

Write a program to figure how thick the paper will be after *n* folds. Assume the unfolded thickness of the paper to be .01 inches.

Be sure to convert inches to feet when you exceed twelve and then feet to miles when you exceed 5280 feet. The output should include, in tabular form, the number of folds and the thickness. Take a guess as to how many folds it takes to produce a piece of paper one mile thick.

The geometric series whose general term is 2^n might prove helpful. Compare the results with the handbook values for successive powers of two.

Consider starting with a piece of paper which has the same area as a football field. When you've figured out how many folds are needed to produce a mile-high pile, try to compute the size of each piece in the pile. It might be better to pose this part of the problem by allowing the paper to be cut and stacked. It is difficult to fold a piece of paper more than 8 or 10 times.

FOLDS	THICKNESS
*****	*********
0	.01 INCHES
1	.02 INCHES
2	.04 INCHES
3	.08 INCHES
4	.16 INCHES
5	.32 INCHES
6	.64 INCHES
7	1.28 INCHES
8	2.56001 INCHES
9	5.12002 INCHES
10	10.24 INCHES
11	1.70667 FEET
12	3.41335 FEET
13	6.82669 FEET
14	13.6534 FEET
15	27.3068 FEET
16	54.6136 FEET
17	109.227 FEET
18	218.455 FEET
19	436.91 FEET
20	873.82 FEET
21	1747.64 FEET
22	3495.28 FEET
23	1.32397 MILES

```
10 PRINT 'FOLDS','THICKNESS'
20 PRINT '*****','*********'
30 PRINT
40 FOR X = 0 TO 30
50 LET Y = .01*(2↑X)
60 IF Y > = 12 THEN 75
70 PRINT X,Y;' INCHES'
72 GO TO 150
75 IF Y/12 > = 5280 THEN 90
80 PRINT X,Y/12;' FEET'
85 GO TO 150
90 PRINT X,Y/(5280*12);' MILES'
150 NEXT X
200 END
```

ANALYSIS

The program is elementary and within the capability of any beginning programmer. Notice line 50 takes into account the thickness of a single piece of paper. Here it is assumed that the paper is .01 inches thick. An interesting exercise would be to have the student measure, somehow, the thickness of a single sheet to avoid compounding of errors.

The significance of successive doubling becomes obvious in this program and the ones which follow. Most students would never have guessed that it takes only 23 folds for the paper to be a mile high. Interesting discussions usually can be made with the computer output from any such program.

One discussion might involve the error introduced by approximate measurements of the thickness of the paper. Another discussion might involve the physical size of the piece of paper necessary to achieve 23 folds. Computing its area might be interesting. How much might it weigh and how many men would it take to fold it?

Suppose a man came to work for you. He didn't want to be paid like everyone else. He asked for a special system of payment based on doubling.

He wanted to be paid twice a month for a total of 24 pay periods per year. He wanted only 1¢ for his first pay period, 2¢ for his second pay period, 4¢ for his third and so on, each time doubling his previous pay-check, if his work was deemed satisfactory.

As an intelligent employer, you should first like to investigate the financial ramifications of this.

Program your computer to print out the man's salary for each pay period along with a running total of how much he has received to date.

PAY PERIOD	SALARY$$	SALARY TO DATE
JAN 1	$ .01	$ .01
JAN 2	$ .02	$ .03
FEB 1	$ .04	$ .07
FEB 2	$ .08	$ .15
MAR 1	$ .16	$ .31
MAR 2	$ .32	$ .63
APR 1	$ .64	$ 1.27
APR 2	$ 1.28	$ 2.550
MAY 1	$ 2.56	$ 5.110
MAY 2	$ 5.12	$ 10.230
JUN 1	$ 10.24	$ 20.470
JUN 2	$ 20.48	$ 40.950
JUL 1	$ 40.96	$ 81.910
JUL 2	$ 81.92	$ 163.830
AUG 1	$ 163.84	$ 327.670
AUG 2	$ 327.68	$ 655.350
SEP 1	$ 655.36	$ 1310.710
SEP 2	$ 1310.72	$ 2621.430
OCT 1	$ 2621.44	$ 5242.870
OCT 2	$ 5242.88	$ 10485.75
NOV 1	$ 10485.76	$ 20971.51
NOV 2	$ 20971.52	$ 41943.030
DEC 1	$ 41943.04	$ 83886.069
DEC 2	$ 83886.08	$ 167772.15

```
10 PRINT 'PAY PERIOD','SALARY$$','SALARY TO DATE'
20 PRINT
30 LET T=0
40 FOR N=0 TO 23
50 READ A$
60 LET S=(2↑N)*.01
70 LET T=T+S
80 PRINT A$,'$ ';S,'$ ';T
90 NEXT N
95 DATA JAN 1,JAN 2,FEB 1,FEB 2,MAR 1,MAR 2,APR 1,APR 2,MAY 1
96 DATA MAY 2,JUN 1,JUN 2,JUL 1,JUL 2,AUG 1,AUG 2,SEP 1,SEP 2
97 DATA OCT 1,OCT 2,NOV 1,NOV 2,DEC 1,DEC 2
100 END
```

4 The First Big Deal Involving Wheat

The story is told of an ancient kingdom with a lazy king. Though benevolent and well-liked he was extremely fond of fun and frolic. He had many jesters and magicians in his court.

He came to enjoy many things. At the hands of a particularly gifted mathematician he came to enjoy the intricacies of mathematics. The mathematician taught him many tricks and games. The king quickly mastered them all. He commissioned the man to come up with a new and challenging game whereupon the magician invented CHESS.

The king was delighted, fascinated and anxious to show his gratitude. He offered the mathematician anything in the kingdom. The mathematician declined. The king insisted he take something. The mathematician gave in. He said all he wanted was some grains of wheat for as many days as there were squares on the board. He wanted the amount of wheat computed as follows: a single grain on the first square, two grains on the second square, four grains on the third square, and so on, each time doubling the amount found on the previous square. The legend relates that at first the king thought it a meager request for so great a game as CHESS. He soon came to realize the enormous and impossible order he had to fill. The legend ends with the beheading of the mathematician.

Write a program to compute how many grains of wheat were on each of the 64 squares of the CHESS board. Also have the program print out a running total of the partial sums. Compare the total amount with the total production per year. Some say there is enough wheat involved to cover the entire state of California to a depth of three feet. It is left as an exercise for the student to verify this conjecture.

SQUARE	GRAINS	TOTAL
1	1	1
2	2	3
3	4	7
4	8	15
5	16	31
6	32	63
7	64	127
8	128	255
9	256.001	511.001
10	512.002	1023
11	1024	2047.01
12	2048.01	4095.01
13	4096.02	8191.04
14	8192.03	16383.1
15	16384.1	32767.1
16	32768.2	65535.3
17	65536.3	131072
18	131073	262144
19	262146	524290
20	524292	1.04858E+06
21	1.04858E+06	2.09717E+06
22	2.09717E+06	4.19434E+06
23	4.19434E+06	8.38868E+06
24	8.38868E+06	1.67774E+07
25	1.67774E+07	3.35547E+07
26	3.35547E+07	6.71095E+07
27	6.71094E+07	1.34219E+08
28	1.34219E+08	2.68438E+08
29	2.68438E+08	5.36876E+08
30	5.36877E+08	1.07375E+09
31	1.07375E+09	2.14751E+09
32	2.14751E+09	4.29502E+09
33	4.29501E+09	8.59003E+09
34	8.59003E+09	1.71801E+10
35	1.71801E+10	3.43601E+10
36	3.43601E+10	6.87202E+10
37	6.87206E+10	1.37441E+11
38	1.37440E+11	2.74881E+11
39	2.74882E+11	5.49763E+11
40	5.49765E+11	1.09953E+12
41	1.09953E+12	2.19906E+12
42	2.19906E+12	4.39811E+12
43	4.39812E+12	8.79623E+12
44	8.79623E+12	1.75925E+13
45	1.75925E+13	3.51849E+13

SQUARE	GRAINS	TOTAL
46	3.51849E+13	7.03699E+13
47	7.03699E+13	1.40740E+14
48	1.40740E+14	2.81479E+14
49	2.81481E+14	5.62960E+14
50	5.62962E+14	1.12592E+15
51	1.12592E+15	2.25184E+15
52	2.25184E+15	4.50368E+15
53	4.50367E+15	9.00735E+15
54	9.00739E+15	1.80147E+16
55	1.80148E+16	3.60295E+16
56	3.60295E+16	7.20590E+16
57	7.20587E+16	1.44118E+17
58	1.44117E+17	2.88235E+17
59	2.88236E+17	5.76472E+17
60	5.76473E+17	1.15294E+18
61	1.15295E+18	2.30589E+18
62	2.30589E+18	4.61178E+18
63	4.61181E+18	9.22359E+18
64	9.22361E+18	1.84472E+19

```
10 PRINT 'SQUARE','GRAINS','TOTAL'
20 PRINT
25 LET S=0
30 FOR X=0 TO 63
35 LET S=S+2↑X
40 PRINT X+1,2↑X,S
50 NEXT X
60 END
```

ANALYSIS

The program is really rather trivial. Statement 25 resets the partial sum incrementer to zero for each run. Statement 35 does the successive doubling. Statement 40 prints the exponent, term and partial sum.

The output exceeds six places on the 21st square. An ideal exercise is to rewrite the program to avoid truncation and type out the values with all significant places retained.

Notice the error introduced by the internal conversion from binary to decimal on square 9. This is done automatically by the computer. The program could be made to remove these extraneous places.

5   You Be The Computer

Suppose you find the following program lying around.
Can you determine what it does to the variable *n*?

```
100 PRINT 'NUMBER'
110 INPUT N
120 FOR I = 1 TO N
130 LET A = I*(I + 1)
140 IF A = N THEN 200
150 IF A > N THEN 300
160 NEXT I
170 PRINT
200 PRINT 'ANSWER = ';I
210 GO TO 100
220 PRINT
300 PRINT 'NOT POSSIBLE'
310 GO TO 100
900 END
```

```
NUMBER
? 12
ANSWER =  3
NUMBER
? 34
NOT POSSIBLE
NUMBER
? 45
NOT POSSIBLE
NUMBER
? 23
NOT POSSIBLE
NUMBER
? 20
ANSWER =  4
NUMBER
? 68
NOT POSSIBLE
NUMBER
? 90
ANSWER =  9
NUMBER
? STOP
 PROGRAM STOPPED.
```

ANALYSIS

The program simply takes a number, factors it and if it can
find two consecutive integers as factors, it prints out the
lowest only. If it cannot find two consecutive integers as factors, it prints out *NOT POSSIBLE*.

Actually, the program only investigates consecutive integers. It just keeps multiplying consecutive pairs of integers in
line 130 and tests their product against the number inputed.
This is done in line 140. As soon as the product becomes larger
than the number inputed—line 150—The *NOT POSSIBLE*
message is printed. Where might such a program be useful?

11

1 1 2 3 5 8 13 21 34 55 89 144 233 377 610....

Recognize the sequence of numbers listed above. It's a Fibonacci sequence. It has a number of interesting properties.

Every term is the sum of the two preceding it.

The product of any two adjacent terms is either one more or one less than the two which sandwich it.

The square of any term when added to the previous term is a Fibonacci number. (A number in the original sequence.)

How does this relate to the Golden Ratio? You might first ask what is the Golden Ratio? The Greeks used the ratio 1.618 to 1 as the basis for their architecture. Actually 1.618 is an approximation for $(1 + \sqrt{5})/2$.

Now, how are the two seemingly unrelated concepts brought together? Well, when any term, or rather each sucessive term of the sequence is multiplied by the Golden Ratio (1.618) the product gets successively closer to the next term.

The chart below illustrates what I mean

$$1 \times 1.618 ... = 1.618 = 2 - .392$$
$$2 \times 1.618 ... = 3.236 = 3 + .236$$
$$3 \times 1.618 ... = 4.854 = 5 - .146$$
$$5 \times 1.618 ... = 8.090 = 8 + .090$$
$$8 \times 1.618 ... = 12.944 = 13 - .056$$

and we see that the deviation is decreasing.

Write a program to prove this contention true by continuing the chart until it is no longer feasible to do so.

References:

J.S. Meyer, **More Fun With Mathematics,** pp. 51 ff.

J.H. Caldwell, **Topics In Recreational Math,** pp. 12–20

FIBONACCI AND THE GOLDEN RATIO

```
 1 * 1.61803 .... = 1.61803 = 2 - .381966
 2 * 1.61803 .... = 3.23607 = 3 + .236068
 3 * 1.61803 .... = 4.8541 = 5 - .145898
 5 * 1.61803 .... = 8.09017 = 8 + 9.01699E-02
 8 * 1.61803 .... = 12.9443 = 13 - .055727
 13 * 1.61803 .... = 21.0344 = 21 + 3.44429E-02
 21 * 1.61803 .... = 33.9787 = 34 - .021286
 34 * 1.61803 .... = 55.0132 = 55 + 1.31607E-02
 55 * 1.61803 .... = 88.9919 = 89 - 8.11768E-03
 89 * 1.61803 .... = 144.005 = 144 + 5.03540E-03
 144 * 1.61803 .... = 232.997 = 233 - 3.08228E-03
 233 * 1.61803 .... = 377.002 = 377 + 1.95313E-03
 377 * 1.61803 .... = 609.999 = 610 - 1.09863E-03
 610 * 1.61803 .... = 987.001 = 987 + 8.54492E-04
 987 * 1.61803 .... = 1597 = 1597 - 2.44141E-04
 1597 * 1.61803 .... = 2584 = 2584 + 4.88281E-04
DEVIATION GONE ... TO 6 PLACES!!
```

```
05 PRINT 'FIBONACCI AND THE GOLDEN RATIO'
07 PRINT
10 READ A,C
20 LET G=(1+SQR(5))/2
30 LET B=A*G
40 IF B>=C THEN 80
50 LET D=C-B
60 IF D=0 THEN 115
70 PRINT A;' * ';G;' .... = ';B;' = ';C;' - ';D
75 GO TO 100
80 LET D=B-C
85 IF D=0 THEN 115
90 PRINT A;' * ';G;' .... = ';B;' = ';C;' + ';D
100 LET A1=A
105 LET A=C
107 LET C=A1+C
110 GO TO 20
115 PRINT 'DEVIATION GONE ... TO 8 PLACES!!'
120 DATA 1,2
200 END
```

ANALYSIS

Line 20 utilizes the formula which gives the Golden Ratio its value to eight places. Line 10 is used to read the first two Fibonacci numbers. It's easier to get the sequence started this way than to work with the leading zero.

Line 30 multiplies by the current term. Two routines are then used. One if the product exceeds the next Fibonacci number, another if it is lower than the next term. D is the deviation and is computed in line 50.

The next term is computed in line 107. The current term is stored in A in line 105.

7 Armstrong Numbers

An *n*-digit number is an Armstrong number if the sum of the *n*-th power of the digits is equal to the original number.

For example, 371 is an Armstrong number because it has three digits such that:

$$3^3 + 7^3 + 1^3$$

Write a program to find all Armstrong numbers with 2, 3 or 4 digits.

NOTE: When the number has four digits the fourth power is used. Do a little preliminary research here in number theory to give yourself an idea of how many numbers there are.

```
153 IS AN ARMSTRONG NUMBER
370 IS AN ARMSTRONG NUMBER
371 IS AN ARMSTRONG NUMBER
407 IS AN ARMSTRONG NUMBER
```

```
100 FOR X = 100 TO 999
110 LET A = INT(X/100)
120 LET B = INT(.1*(X-(100*A)))
130 LET C = X-(100*A + 10*B)
140 LET Y = A↑3 + B↑3 + C↑3
150 IF ABS(Y-X) > .1 THEN *+2
160 PRINT X; ' IS AN ARMSTRONG NUMBER'
170 NEXT X
180 PRINT
190 FOR X = 1000 TO 9999
200 LET A = INT(X/1000)
210 LET B = INT(10*FRP(X/1000))
220 LET C = INT(10*FRP(X/100))
230 LET D = INT(10*FRP(X/10))
240 IF A↑4 + B↑4 + C↑4 + D↑4 < > X THEN *+3
250 PRINT
260 PRINT X; ' IS AN ARMSTRONG NUMBER'
270 NEXT X
280 END
```

References:
Back issues of **The Mathematics Teacher.**

ANALYSIS

The digits of the number are separated in lines 110 through 130. The reader is invited to follow a sample calculation along to observe the technique (see *Programming Tricks*, Number 7 in Appendix). Each digit is cubed in line 140.

The output indicates that there are four Armstrong numbers with three digits. The computer found none with four digits. The search could have continued up to five if the other loops had been eliminated.

8 The Famous Indian Problem

Write a program to help the Indians out. The Indians in question are the ones who sold Manhattan Island to the Dutch for the paltry sum of $24.00.

The sale took place in the year 1626. Suppose they had deposited the money in the local bank. Interest rates changed each century according to the table given below.

1600's	2%
1700's	3%
1800's	4%
1900's	5%

Figure out how much the Indians would have in the bank today. Compound the interest annually.

Reference:

Samuel Selby, **CRC Standard Mathematical Tables.**

```
THE FAMOUS INDIAN PROBLEM!!!
THE INDIANS HAVE   3729818.2 DOLLARS ON ACCOUNT!!!
```

```
05 PRINT 'THE FAMOUS INDIAN PROBLEM!!!'
10 LET Y = 24
20 FOR X = 1626 TO 1973
25 IF X > = 1700 THEN 40
30 LET R = .02
35 GO TO 100
40 IF X > = 1800 THEN 55
45 LET R = .03
50 GO TO 100
55 IF X > = 1900 THEN 70
60 LET R = .04
65 GO TO 100
70 LET R = .05
100 LET I = Y*R
110 LET Y = Y + I
120 NEXT X
130 PRINT 'THE INDIANS HAVE ';Y;' DOLLARS ON ACCOUNT!!!'
200 END
```

ANALYSIS

It is obvious that the X loop represents the year in question. Line 10 is the starting account balance. The variable representing the running account balance is Y. R changes from century to century and is the interest rate.

Lines 100 and 110 are the meat of the program. Line 100 computes the interest. Line 110 adds the interest to the previous balance.

Write a program that will read a message from a DATA statement in Morse Code and translate it into English.

Have the computer type out the original coded message and the decoded message beneath it.

Use the following alphanumeric code:

> . = DOT
> - = DASH

Find an old Boy Scout Handbook for the intricacies of the code.

NOTE: Be sure to define and use an end of message character. You might also consider some uniform spacing between each character. Remember the computer is not capable of the fine lines of time interpretation that human beings are.

References:
An interesting device which can be built to do this is described in the March, 1979 issue of **Popular Electronics.**

10   Square Root By Iteration

Write a program to compute the square root of any number. Use the iterative algorithm whereby a guess is entered along with the number to be square-rooted. Have the computer refine that guess by division and averaging until the square root is accurate to six decimal places.

Place a counter in the iterative loop to determine how many iterations were required.

Compare the iterative square root to the functional square root available in BASIC.

Extra points for algorithms other than the successive division one. There are some fascinating geometric and arithmetic methods of square root computation.

References:
Peter Calingaert, **Principles of Computation,** p. 133, 154.
Ladis D. Kovach, **Computer-Oriented Math,** pp. 50–55.

```
SQUARE ROOTS BY ITERATION

TYPE IN THE NUMBER YOU WISH SQUARE ROOTED
FOLLOWED BY YOUR GUESS!!
? 23,5

AFTER  4 ITERATIONS, I FIND THE SQUARE ROOT OF  23
TO BE  4.7958315
THE FUNCTIONAL SQUARE ROOT IS  4.7958315

ANOTHER NO. PLEASE, FOLLOWED BY YOUR GUESS!
? 34,2

AFTER  6 ITERATIONS, I FIND THE SQUARE ROOT OF  34
TO BE  5.8309519
THE FUNCTIONAL SQUARE ROOT IS  5.8309519

ANOTHER NO. PLEASE, FOLLOWED BY YOUR GUESS!
? 33333,7

AFTER  9 ITERATIONS, I FIND THE SQUARE ROOT OF  33333
TO BE  182.57327
THE FUNCTIONAL SQUARE ROOT IS  182.57327

ANOTHER NO. PLEASE, FOLLOWED BY YOUR GUESS!
? STOP
 PROGRAM STOPPED.
```

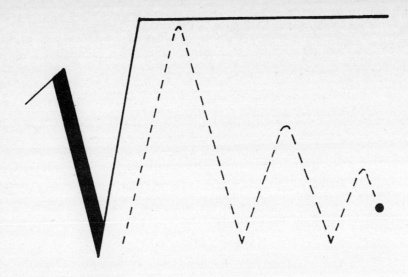

```
10 PRINT 'SQUARE ROOTS BY ITERATION'
20 PRINT
25 PRINT 'TYPE IN THE NUMBER YOU WISH SQUARE ROOTED '
30 PRINT 'FOLLOWED BY YOUR GUESS!!'
35 GO TO 50
40 PRINT
45 PRINT 'ANOTHER NO. PLEASE, FOLLOWED BY YOUR GUESS!'
50 INPUT B,A
53 LET G = 0
54 IF A = 0 THEN PRINT 'HEY FELLA, IT SMARTS TO DIVIDE BY 0!'
55 FOR C = 1 TO 1000
60 LET D = B/A
65 LET E = (D + A)/2
70 LET G = G + 1
75 IF A = E THEN 100
80 LET A = E
85 NEXT C
100 PRINT
110 PRINT 'AFTER ';G;' ITERATIONS, I FIND THE SQUARE ROOT OF ';B
115 PRINT 'TO BE ';E
120 PRINT 'THE FUNCTIONAL SQUARE ROOT IS ';SQR(B)
130 GO TO 40
200 END
```

ANALYSIS

This program is a good example of the binary search routine, one of the most efficient subroutines available for iterative procedures. It simple means you halve the interval you're working with to iterate to the next step.

In this program the number whose square root we seek is divided by a guess (line 60), the quotient and the divisor are averaged to establish the next guess (lines 65 and 80). Line 75 tests the two to see if they are the same. When they are the division will have produced the desired square root. The iterations are counted by the variable G. The reader is invited to try a similar procedure for roots other than 2.

11   Roman Numeral Addition And Multiplication

Write a program to add and multiply any two Roman numerals inputed.

Perform both operations on the numbers and print both answers in Roman form. Have the computer type out the Arabic equivalent underneath each set of operations.

In other words have the computer state the problem concisely. For example:

$$XVII + XLIV = LXI$$

Don't feel bad here. Archimedes and others were able to do extensive calculations in this system. How thankful they would be if they could work in our modern decimal system.

Separate the function of the conversion from ARABIC to ROMAN and vice-versa from the computation.

```
ROMAN NUMERAL ADDITION AND MULTIPLICATION

TYPE TWO INTEGER NUMBERS, EACH LESS THAN 2000
? 532,5

ADDITION:

DXXXII + V = DXXXVII
( 532  +  5  =  537 )


MULTIPLICATION

DXXXII * V = MMDCLX
( 532  *  5  =  2660 )

ROMAN NUMERAL ADDITION AND MULTIPLICATION

TYPE TWO INTEGER NUMBERS, EACH LESS THAN 2000
? 365,11

ADDITION:

CCCLXV + XI = CCCLXXVI
( 365  +  11  =  376 )


MULTIPLICATION

CCCLXV * XI = MMMMXV
( 365  *  11  =  4015 )
```

```
100 PRINT "ROMAN NUMERAL ADDITION AND MULTIPLICATION"
110 PRINT
120 PRINT "TYPE TWO INTEGER NUMBERS,
    EACH LESS THAN 2000"
130 INPUT A,B
140 IF A>2000 THEN 120
150 IF B>2000 THEN 120
160 PRINT
170 PRINT "ADDITION:"
180 PRINT
190 C=A+B
200 Z=A
210 GOSUB 460
220 PRINT " + ";
230 Z=B
240 GOSUB 460
250 PRINT " = ";
260 Z=C
270 GOSUB 460
280 PRINT
290 PRINT "(";A;" + ";B;" = ";C;")"
300 PRINT
310 PRINT
320 PRINT "MULTIPLICATION"
330 PRINT
340 C=A*B
350 Z=A
360 GOSUB 460
370 PRINT " * ";
380 Z=B
390 GOSUB 460
400 PRINT " = ";
410 Z=C
420 GOSUB 460
430 PRINT
440 PRINT "(";A;" * ";B;" = ";C;")"
450 GOTO 590
460 RESTORE
470 FOR I=1 TO 7
480 READ D
490 N=INT(Z/D)
500 READ D$
510 IF N=0 THEN 560
520 FOR J=1 TO N
530 PRINT D$;
540 NEXT J
550 Z=Z-N*D
560 NEXT I
570 RETURN
580 DATA 1000,"M",500,"D",100,"C",
    50,"L",10,"X",5,"V",1,"I"
590 END
```

```
6 DIM X(65)
7 DIM M(39)
8 MAT READ X
9 PRINT
10 PRINT 'ENTER ROMAN CHARACTERS OR
   ARABIC NUMERALS';
20 INPUT R$
30 Q=L=1
40 CHANGE R$ TO M
50 T=0
70 A=1000
72 IF M(M(0)) > 47 THEN 350
75 Q=0
80 IF M(M(0))=14 THEN180
90 IF M(M(0))=27 THEN 190
100 IF M(M(0))=29 THEN200
110 IF M(M(0))=17 THEN 210
120 IF M(M(0))=8 THEN 220
130 IF M(M(0))=9 THEN 230
140 IF M(M(0))=18 THEN 240
150 M(1)=M(M(0))
152 M(0)=1
154 CHANGE M TO I$
160 PRINT 'IMPROPER
    CHARACTER ';I$
170 GO TO 9
180 A=A-4
190 A=A-5
200 A=A-40
210 A=A-50
220 A=A-400
230 A=A-500
240 IF A<L THEN 310
250 T=T+A
260 L=A
270 M(0)=M(0)-1
280 IF M(0)=0 THEN 330
290 GO TO 70
310 T=T-A
320 GO TO 270
330 PRINT 'THE ARABIC
    EQUIVALENT IS ';T
340 GO TO 9
350 IF Q=0 THEN 150
360 FOR W=1 TO M(0)
380 IF M(W)<48 THEN 150
390 NEXT W
400 R=VAL (R$)
410 IF R > 30000 THEN 530
420 M(0)=0
430 FOR K=5 TO 65 STEP 5
440 IF R< X(K-4) THEN 510
450 M(0)=M(0)+X(K-3)
460 M(M(0))=X(K-2)
470 M(0)=M(0)+X(K-1)
480 M(M(0))=X(K)
490 R=R-X(K-4)
500 GO TO 440
510 IF R=0 THEN 550
520 NEXT K
530 PRINT 'CANNOT
    EASILY COMPUTE'
540 GO TO 9
550 CHANGE M TO I$
560 PRINT 'ROMAN
    EQUIVALENT IS ';I$
570 GO TO 9
580 DATA 1000,1,0,0,18
590 DATA 900,1,8,1,18
600 DATA 500,1,0,0,9
610 DATA 400,1,8,1,9
620 DATA 100,1,0,0,8
630 DATA 90,1,29,1,8
640 DATA 50,1,0,0,17
650 DATA 40,1,29,1,17
660 DATA 10,1,0,0,29
670 DATA 9,1,14,1,29
680 DATA 5,1,0,0,27
690 DATA 4,1,14,1,27
700 DATA 1,1,0,0,14
710 END
```

ANALYSIS

This program has been designed to perform the addition and multiplication in base 10 arithmetic and then convert the three numbers, digit-by-digit to Roman numerals. The repeated conversions provide an excellent illustration of the use of a subroutine.

The reader will notice that no attempt has been made to avoid four, or more, successive appearances of the same letter in the Roman numeral. For example, the output might show CCCC for 400 instead of CD. The reader should find programming the modification necessary to give the proper letter pattern an interesting problem.

A second program is provided which simply converts from Roman to Arabic or vice versa.

Conversion To and From Base 10

Write a program to take any number *n* base 10 and convert it to its equivalent in a given base beginning with base 2. The program should also be capable of converting any number back to its base 10 (decimal) equivalent.

As input the user should include the base in current use, the desired base and perhaps the number of digits in the original number. This will simplify the computation.

Algorithms for this program are numerous. The most popular being retention of the remainders after successive divisions by the desired base. Remember, with this method, the number in question is read by listing the remainders in the reverse order of their generation.

Be careful when working in bases above base 10. Here, symbols should be used which take up only one character. Otherwise, the number 11 for example may be confused with two successive 1's.

When converting to base 10, simply generate successive powers of the base in question and multiply. Then add the results to reconstruct the base 10 equivalent.

```
THIS PROGRAM WILL CONVERT TO AND FROM BASE TEN
TO ANY BASE SPECIFIED


INPUT THE CURRENT BASE, BASE DESIRED AND THE NUMBER
ALSO TYPE HOW MANY DIGITS ARE IN THE NUMBER
? 10,2,114,3

  114 BASE  10 IS AS FOLLOWS IN BASE  2
  1 1 1 0 0 1 0

INPUT THE CURRENT BASE, BASE DESIRED AND THE NUMBER
ALSO TYPE HOW MANY DIGITS ARE IN THE NUMBER
? 2,10,1101011,7

  1101011 BASE  2 IS AS FOLLOWS IN BASE  10
  1101011 BASE  2 IS  107 IN BASE  10


INPUT THE CURRENT BASE, BASE DESIRED AND THE NUMBER
ALSO TYPE HOW MANY DIGITS ARE IN THE NUMBER
? 3,10,34,2

  34 BASE  3 IS AS FOLLOWS IN BASE  10
NUMBER IS ILLEGAL IN BASE  3
```

1011000 = 88

```
05 DIM D(40)
10 PRINT 'THIS PROGRAM WILL CONVERT TO AND FROM BASE TEN'
20 PRINT 'TO ANY BASE SPECIFIED'
22 LET S1 = 0
30 PRINT
35 PRINT
40 PRINT 'INPUT THE CURRENT BASE, BASE DESIRED AND THE NUMBER'
45 PRINT 'ALSO TYPE HOW MANY DIGITS ARE IN THE NUMBER'
50 INPUT A,B,C,N
55 LET W = C
60 PRINT
65 PRINT C;' BASE ';A;' IS AS FOLLOWS IN BASE ';B
70 IF A < > 10 THEN 200
80 LET K = 1
90 LET D(K) = C-(INT(C/B)*B)
110 LET K = K + 1
120 LET C = INT(C/B)
130 IF C = 0 THEN 400
140 GO TO 90
200 FOR E = N-1 TO 0 STEP -1
210 LET I = INT(C/10↑E)
215 IF I > = A THEN 460
220 LET R = C-I*10↑E
230 LET C = R
240 LET S = I*(A↑E)
250 LET S1 = S1 + S
260 NEXT E
300 PRINT W;' BASE ';A;' IS ';S1;' IN BASE ';B
310 GO TO 30
400 FOR K = 35 TO 1 STEP -1
410 IF T = 7 THEN 430
420 IF D(K) = 0 THEN 450
430 PRINT D(K);
440 LET T = 7
450 NEXT K
455 GO TO 30
460 PRINT 'NUMBER IS ILLEGAL IN BASE ';A
500 END
```

ANALYSIS

The program was limited to base 10 because of the length of the procedure for inter-base conversion. Actually most of the calculations involve base 10 anyway. The only other practical application that couldn't be done directly would be conversion from base 2 to base 8, that is binary to octal and vice-versa.

The meat of the program is line 90. It stores the remainders after division by the base B because they are usually printed out in reverse order. The statement could have been shortened by using the MOD function which returns the remainder after division by a variable. The quotient is stored in the variable C and is made smaller by the successive division method. When C is zero the process is complete and the program branches to 400 to print the result. It is assumed that the number to be printed will not contain more than 35 digits. Statement 420 suppresses the printing of leading zeroes.

Line 215 insures that such things as 110201 in base 2 are not computed. It checks to see that no number in the base in question either equals or exceeds the base. The subroutine in 220 converts back to base 10 by simply computing successive powers of the base and summing up.

23

13   G.C.D. and L.C.M.

Write a program to fine the G.C.D. (greatest common divisor) and the L.C.M. (least common multiple)—sometimes called the lowest common denominator of a set of numbers.

As input allow up to ten numbers, have the computer factor the numbers and then use a famous algorithm—you find it—to develop and print the G.C.D. and L.C.M.

This problem is not difficult; so try to meet the challenge that faces every programmer—the length of the program.

Try to make the program as concise as it is efficient.

Euclid has all the hints you'll need for this one.

$$2 \times 3 \times 5 \times 7 = 210$$

```
HOW MANY NUMBERS? 5
TYPE IN THE NUMBERS FROM

? 6
? 12
? 13
? 15
? 36
  6 HAS PRIME FACTORS 2   3
  12 HAS PRIME FACTORS 2   2   3
  13 IS PRIME
  15 HAS PRIME FACTORS 3   5
  36 HAS PRIME FACTORS 2   2   3   3

G.C.D. IS 1

L.C.M. IS  2340
ANYMORE TO DO (Y OR N)? Y
```

```
HOW MANY NUMBERS? 3
TYPE IN THE NUMBERS FROM
 LOWEST TO HIGHEST
? 6
? 12
? 21
  6 HAS PRIME FACTORS 2   3
  12 HAS PRIME FACTORS 2   2   3
  21 HAS PRIME FACTORS 3   7

G.C.D. IS 3

L.C.M. IS 84
ANYMORE TO DO (Y OR N)? Y
```

```
HOW MANY NUMBERS? 3
TYPE IN THE NUMBERS FROM
 LOWEST TO HIGHEST
? 12
? 24
? 48
  12 HAS PRIME FACTORS 2   2   3
  24 HAS PRIME FACTORS 2   2   2   3
  48 HAS PRIME FACTORS 2   2   2   2   3

G.C.D. IS 12

L.C.M. IS  48
ANYMORE TO DO (Y OR N)? N
```

References:

Robert Wisner, **A Panorama of Numbers,** p. 142–152.

Edwin Stein, **Fundamentals of Mathematics.**

```
05 DIM B(10)
10 DIM A(10),A$(10)
15 PRINT
20 PRINT 'HOW MANY NUMBERS';
25 INPUT N
30 PRINT 'TYPE IN THE NUMBERS FROM LOWEST TO HIGHEST'
35 FOR X = 1 TO N
40 INPUT A(X)
45 NEXT X
50 FOR X = 1 TO N
55 LET C = 0
62 LET B(X) = A(X)
65 FOR F = 2 TO A(X)
67 IF F = B(X) THEN 120
70 IF A(X)/F < > INT(A(X)/F) THEN 110
72 IF C = 0 THEN PRINT A(X);' HAS PRIME FACTORS ';
75 PRINT F;
80 LET C = C + 1
100 LET A(X) = A(X)/F
105 GO TO 65
110 NEXT F
112 PRINT
115 IF C > 0 THEN 130
120 PRINT A(X);' IS PRIME'
130 NEXT X
140 FOR Z = 2 TO B(N)
142 LET Y = 0
145 FOR X = 1 TO N
150 IF B(X)/Z < > INT(B(X)/Z) THEN 200
165 LET Y = Y + 1
200 NEXT X
210 IF Y < N THEN 250
220 LET T = Z
250 NEXT Z
255 PRINT
260 PRINT 'G. C. D. IS ';T
270 PRINT
280 LET D = B(N)
290 FOR E = 1 TO N
300 IF D/B(E) < > INT(D/B(E)) THEN 330
320 NEXT E
325 PRINT 'L. C. M. IS ';D
327 GO TO 400
330 LET D = D + 1
340 GO TO 290
400 PRINT 'ANYMORE TO DO (Y OR N)';
410 INPUT A$
420 IF A$ = 'Y' THEN 15
500 END
```

ANALYSIS

This program is a bit more complex than a simple divisor generator. We must print out factors here.

This is done in the F loop from 65 to 110. When a number is divisible by F, F is printed and the quotient is recycled to be factored again.

The numbers are entered in a loop to make the program interactive. Line 100 recycles the quotient to be factored again. Should none of the numbers in F be divisors then the variable C in line 80 will remain at zero and the number will be prime. This is tested for in lines 115 and 120.

The G.C.D. is obtained by trying all numbers up to the least number in the list. When all numbers are successfully divided the result is stored in T (line 220), the last such T is the G.C.D. When the incremented variable Y is as big as N then all numbers have been successfully divided by Z.

A similar procedure is used for the L.C.M. Here we start with the largest listed number and increment upward by 1 until we find a number into which all the others go evenly.

The program is limited to 10 numbers.

X	Y	GCD
12	14	2
45	72	9
112	144	16
15	90	15

OUT OF ARITHMETIC DATA IN 20.

RUN STOPPED.

```
10 PRINT ' X',' Y','GCD'
20 READ X,Y
25 PRINT X,Y;
30 LET Q = INT(X/Y)
40 LET R = X-Q*Y
50 LET X = Y
60 LET Y = R
70 IF R > 0 THEN 30
80 PRINT ,X
90 GO TO 20
100 DATA 12,14,45,72,112,144,15,90
200 END
```

ANALYSIS

The Euclidian Algorithm for finding the G.C.D. of a given set of numbers is provided to show the contrast in program length with the previous algorithm.

The procedure itself is classic and known to most teachers. Even the computer programming novice should be able to follow the program.

The reader is invited to modify this program so that it will do exactly what the previous program does.

Write a program to solve a system of two equations in two unknowns.

Input the coefficients as given in the following scheme:

$$a x + b y = c$$
$$d x + e y = f$$

Have the program test for a solution first. If it is determined that a solution exists print it out. You may use one of many algorithms.

Useful algorithms exist in the following areas: determinants, matrices, graphing, slopes or substitution.

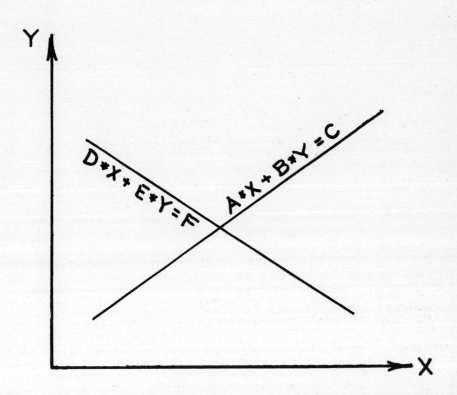

```
SOLUTION OF SIMULTANEOUS EQUATIONS
USING CRAMERS RULE AND DETERMINANTS

ARE YOUR EQUATIONS IN THE RIGHT FORMAT(Y OR N)? Y
TYPE IN THE COEFFICIENTS FROM LEFT TO RIGHT
INCLUDING CONSTANTS ON THE RIGHT HAND SIDE
? 1,1,11,1,-1,5
SOLUTION HAS X = 8 AND Y = 3

ANY MORE TO DO (Y OR N)
? Y

ARE YOUR EQUATIONS IN THE RIGHT FORMAT(Y OR N)? Y
TYPE IN THE COEFFICIENTS FROM LEFT TO RIGHT
INCLUDING CONSTANTS ON THE RIGHT HAND SIDE
? 1,2,16,1,2,16
YOUR EQUATIONS HAVE NO SOLUTION!!!
EVEN I CANT HELP YOU
ANY MORE TO DO (Y OR N)
? N
```

```
10 PRINT
20 PRINT 'SOLUTION OF SIMULTANEOUS EQUATIONS'
25 PRINT 'USING CRAMERS RULE AND DETERMINANTS'
30 PRINT
35 PRINT 'ARE YOUR EQUATIONS IN THE RIGHT FORMAT';
40 PRINT '(Y OR N)';
45 INPUT A$
50 IF A$ = 'Y' THEN 70
55 PRINT 'WELL HOP TO IT....AND TRY ME LATER!!'
60 STOP
70 PRINT 'TYPE IN THE COEFFICIENTS FROM LEFT TO RIGHT'
80 PRINT 'INCLUDING CONSTANTS ON THE RIGHT HAND SIDE'
85 INPUT A,B,C,D,E,F
90 IF A*E-B*D = 0 THEN 200
95 LET X = (C*E-B*F)/(A*E-B*D)
100 LET Y = (A*F-C*D)/(A*E-B*D)
110 PRINT 'SOLUTION HAS X = ';X;' AND Y = ';Y
120 PRINT
130 PRINT 'ANY MORE TO DO (Y OR N)'
140 INPUT B$
150 IF B$ = 'Y' THEN 30
160 STOP
200 PRINT 'YOUR EQUATIONS HAVE NO SOLUTION!!!'
210 PRINT 'EVEN I CANT HELP YOU'
220 GO TO 130
300 END
```

ANALYSIS

Cramer's Rule for systems of equations works whenever the coefficient matrix for the system has a non-zero determinant.

The algorithm simply computes two determinants and divides them to establish the value of each variable. The reader can follow the program in lines 95 to 100 to see which variables are involved. The test for a unique solution is in line 90.

A detailed discussion of the theory is either already known by most teachers or can be found in any linear algebra or intermediate mathematics text.

Write a program to calculate the first 50 terms of the geometric sequence given below. Use the three sets of parameters listed.

$$B + BX + BX^2 + BX^3 + \ldots + BX^{49}$$

Compare your sum with the sum given by the formula for the first n terms of a geometric series.

Use the following parameters:
1) $B = 1$ and $X = 1/2$
2) $B = 4$ and $X = 1/2$
3) $B = 2$ and $X = 1/4$

HINT:

A loop with an accumulator is called for here. It might look like this in BASIC:

$$65 \text{ LET } M = M + Z$$

where Z is the term under computation and M is the partial sum.

Expand your program to accept any geometric or arithmetic series. The program could evaluate sums, individual terms and print out the series.

Reference:

Donald Greenspan, **Introduction to Calculus**, pp. 138–42.

```
THIS PROGRAM WILL COMPUTE THE N TH TERM AND SUM
OF AN ARITHMETIC OR GEOMETRIC SEQUENCE.

IS THIS SEQUENCE ARITHMETIC OR GEOMETRIC, TYPE(A OR G)
? A
INPUT FIRST TERM, COMMON DIFFERENCE & NO. OF TERMS
? 3,4,15
 15 TH TERM IS  59
SUM OF  15 TERMS IS  465

DO YOU WISH SEQUENCE LISTED ( Y OR N)? Y
 3 7 11 15 19 23 27 31 35 39 43 47 51 55 59

ANY MORE TO DO (Y OR N)? Y

IS THIS SEQUENCE ARITHMETIC OR GEOMETRIC, TYPE(A OR G)
? G
WHAT IS THE FIRST TERM, COMMON RATIO & NO. OF TERMS
? 4,2,13

 13 TH TERM IS  16384
SUM OF  13 TERMS IS  32764

DO YOU WISH THE SEQUENCE LISTED, (Y OR N)? Y
 4 8 16 32 64 128 256 512 1024 2048 4096 8192 16384

ANY MORE TO DO (Y OR N)? N
```

```
10 DIM A$(10),B$(10),C$(10),D$(10)
20 PRINT 'THIS PROGRAM WILL COMPUTE THE N TH TERM AND SUM'
22 PRINT 'OF AN ARITHMETIC OR GEOMETRIC SEQUENCE.'
25 PRINT
30 PRINT 'IS THIS SEQUENCE ARITHMETIC OR GEOMETRIC, TYPE(A OR G)'
35 INPUT A$
40 IF A$ = 'G' THEN 100
45 PRINT 'INPUT FIRST TERM, COMMON DIFFERENCE & NO. OF TERMS'
50 INPUT A,D,N
55 LET L = A + (N-1)*D
60 PRINT N;' TH TERM IS ';L
65 PRINT 'SUM OF ';N;' TERMS IS ';(N/2)*(A + L)
70 PRINT
72 PRINT 'DO YOU WISH SEQUENCE LISTED ( Y OR N)';
75 INPUT D$
80 IF D$ = 'N' THEN 200
85 FOR X = 1 TO N
90 PRINT A;
95 LET A = A + D
97 NEXT X
99 GO TO 200
100 PRINT 'WHAT IS THE FIRST TERM, COMMON RATIO & NO. OF TERMS'
110 INPUT A,R,N
120 PRINT
125 PRINT N;' TH TERM IS ';A*(R↑(N-1))
130 PRINT 'SUM OF ';N;' TERMS IS ';(A*(R↑N-1))/(R-1)
140 PRINT
145 PRINT 'DO YOU WISH THE SEQUENCE LISTED, (Y OR N)';
150 INPUT B$
160 IF B$ = 'N' THEN 200
165 FOR X = 1 TO N
167 PRINT A;
170 LET A = A*R
175 NEXT X
200 PRINT
202 PRINT
205 PRINT 'ANY MORE TO DO (Y OR N)';
210 INPUT C$
220 IF C$ = 'Y' THEN 25
300 END
```

ANALYSIS

The program simply uses the well-known formulas for arithmetic and geometric sequences.

If the sequence is arithmetic the information is collected in line 50. The last term is computed in line 55 and the sum in line 65 which also prints the sum. If the sequence is geometric the computations take place in lines 125 and 130.

The problem is solved simply by adding the common difference or ratio to the last term as many times as there are terms to be printed. This is done in lines 85 through 97 for arithmetic, and lines 165 through 175 for geometric.

16 Solution of a Quadratic Equation

Write a program to solve a quadratic equation of the form:

$$ax^2 + bx + c = 0$$

when a, b and c are inputed.

Be sure to test for a negative discriminant and print an appropriate warning. Also be sure to include a test for the zero denominator before the computer divides by that zero.

Use the quadratic formula to predict the exact nature of the roots before they are actually computed.

A little extra effort should enable you to have the program recognize and accomodate imaginary roots and type them out in the form

$$a + bi$$

Expand your program to include the sum and product of the roots.

Perhaps you can work backwards and print out the quadratic equation when given the sum and product of the roots.

```
INPUT COEFFICIENTS OF QUADRATIC IN FORM AX↑2+BX+C=0
? 0,5,6
THE EQUATION IS LINEAR NOT QUADRATIC!!
ANY MORE TO DO (Y OR N)
? Y


INPUT COEFFICIENTS OF QUADRATIC IN FORM AX↑2+BX+C=0
? 1,5,6
ROOTS ARE REAL, RATIONAL AND UNEQUAL
THEY ARE -2    -3
ANY MORE TO DO (Y OR N)
? Y


INPUT COEFFICIENTS OF QUADRATIC IN FORM AX↑2+BX+C=0
? 1,6,9
ROOTS ARE REAL, RATIONAL AND EQUAL
THEY ARE -3    -3
ANY MORE TO DO (Y OR N)
? Y


INPUT COEFFICIENTS OF QUADRATIC IN FORM AX↑2+BX+C=0
? 6,5,7
ROOTS ARE IMAGINARY!!
ROOTS ARE :
-.41666666 +  .99652172 I
-.41666666 -  .99652172 I
ANY MORE TO DO (Y OR N)
? N
```

```
03 DIM A$(10)
05 PRINT
10 PRINT 'INPUT COEFFICIENTS OF QUADRATIC IN FORM AX↑2 + BX + C = 0'
20 INPUT A,B,C
30 IF A = 0 THEN 200
40 LET D = B↑2 - 4*A*C
45 IF D < 0 THEN 250
50 LET X1 = (-B + SQR(D))/(2*A)
60 LET X2 = (-B-SQR(D))/(2*A)
70 IF D = 0 THEN 220
80 IF SQR(D) < > INT(SQR(D)) THEN 110
90 PRINT 'ROOTS ARE REAL, RATIONAL AND UNEQUAL'
100 PRINT 'THEY ARE ';X1,X2
105 GO TO 300
110 PRINT 'ROOTS ARE REAL, IRRATIONAL AND UNEQUAL'
120 PRINT 'THEY ARE APPROXIMATELY ';X1,X2
130 GO TO 300
200 PRINT 'THE EQUATION IS LINEAR NOT QUADRATIC!!'
210 GO TO 300
220 PRINT 'ROOTS ARE REAL, RATIONAL AND EQUAL'
230 PRINT 'THEY ARE ';X1,X2
240 GO TO 300
250 PRINT 'ROOTS ARE IMAGINARY!!'
260 PRINT 'ROOTS ARE :'
265 PRINT -B/(2*A);' + ';SQR(ABS(D))/(2*A);' I'
267 PRINT -B/(2*A);' - ';SQR(ABS(D))/(2*A); ' I'
300 PRINT 'ANY MORE TO DO (Y OR N)'
310 INPUT A$
320 IF A$ = 'Y' THEN 05
400 END
```

ANALYSIS

The quadratic formula has been used. Some frills have been added.

Line 30 checks to be sure the equation is second degree. Line 40 computes the discriminant. Line 45 accesses the subroutine for computing the imaginary roots. This is done in lines 265 and 267. If the roots are real they are computed in lines 50 and 60.

The rest of the program contains simple conditional branches. The discriminant is tested and the computer prints the appropriate messages. Line 80 tests to see if the discriminant is a perfect square. Notice the mathematical techniques for doing that.

This program can handle any input a student may give it. It is always best to try to include these contingencies in the program. That way the error messages typed out will be your own. Some error messages in BASIC are obscure. BASIC is, however, far better than most programming languages as far as understanding error messages is concerned.

Every algebra student has cringed at the thought of doing word problems. All word problems are not alike. Perhaps the most fearsome type are the mixture problems. Some teachers avoid them.

If you can program these type problems then you've got them licked. Write a program to solve mixture problems of the type given below.

Suppose it is desired to mix a certain amount of coffee at 69¢ per pound with 23 pounds of coffee at 98¢ per pound. We want to know how many pounds of the 69¢ coffee we will need to make a mixture which will cost 73¢ per pound.

Let any of the above quantities be available as unknowns. Do the problem in general, program it, and you'll find you've gone a long way towards understanding this type of problem.

```
WHAT UNITS USED (LBS,OZ,ETC) AND DENOMINATION (DOLLARS,CENTS)
? POUNDS,CENTS
INPUT # OF POUNDS AND CENTS PER POUNDS OF FIRST ITEM
IF A QUANTITY IS UNKNOWN, ENTER ZERO
? 0,120
INPUT # OF POUNDS AND CENTS PER POUNDS OF SECOND ITEM
? 0,70
INPUT # OF POUNDS AND CENTS PER POUNDS OF MIXTURE
? 50,100
MIX  30 POUNDS AT  120 CENTS PER POUNDS
WITH  20 POUNDS AT  70 CENTS PER POUNDS
TO GET A MIXTURE OF  50 POUNDS AT  100 CENTS PER POUNDS
ANYTHING ELSE (YES OR NO)? YES


WHAT UNITS USED (LBS,OZ,ETC) AND DENOMINATION (DOLLARS,CENTS)
? BARRELS,DOLLARS
INPUT # OF BARRELS AND DOLLARS PER BARRELS OF FIRST ITEM
IF A QUANTITY IS UNKNOWN, ENTER ZERO
? 0,3.20
INPUT # OF BARRELS AND DOLLARS PER BARRELS OF SECOND ITEM
? 10,5.60
INPUT # OF BARRELS AND DOLLARS PER BARRELS OF MIXTURE
? 0,4
MIX  20 BARRELS AT  3.20 DOLLARS PER BARRELS
WITH  10 BARRELS AT  5.6 DOLLARS PER BARRELS
TO GET A MIXTURE OF  30 BARRELS AT  4 DOLLARS PER BARRELS
ANYTHING ELSE (YES OR NO)? NO
```

```
100 PRINT'WHAT UNITS USED (LBS,OZ,ETC) AND DENOMINATION (DOLLARS,CENTS)'
110 INPUT U$,M$
120 PRINT'INPUT # OF ';U$;' AND ';M$;' PER ';U$;' OF FIRST ITEM'
130 PRINT'IF A QUANTITY IS UNKNOWN, ENTER ZERO'
140 INPUT N1,C1
150 PRINT'INPUT # OF ';U$;' AND ';M$;' PER ';U$;' OF SECOND ITEM'
160 INPUT N2,C2
170 PRINT'INPUT # OF ';U$;' AND ';M$;' PER ';U$;' OF MIXTURE'
180 INPUT N3,C3
190 LET A=(C1+C2)/2
200 IF N1=0 AND N2=0 THEN 310
210 IF N1=0 AND N3=0 THEN 510
220 IF N2=0 AND N3=0 THEN 510
230 IF N1=0 THEN N1=(N3*C3-N2*C2)/C1
240 IF N2=0 THEN N2=(N3*C3-N1*C1)/C2
250 N3=N1+N2
260 IF C3=0 THEN C3=(N1*C1+N2*C2)/N3
270 PRINT'MIX ';N1;' ';U$;' AT ';C1;' ';M$;' PER '; U$
280 PRINT'WITH ';N2;' ';U$;' AT ';C2;' ';M$;' PER ';U$
290 PRINT'TO GET A MIXTURE OF ';N3;' ';U$;' AT ';C3;' ';M$;' PER ';U$
300 GO TO 780
310 IF A>C3 THEN 350
320 LET L1=N3/2
330 LET M1=N3
340 GO TO 370
350 LET L1=0
360 LET M1=N3/2
370 LET K2=N3*C3
380 FOR X=L1 TO M1 STEP .1
390 IF X*C1+(N3-X)*C2=N3*C3 THEN 480
400 LET K1 = N3*C3-X*C1-(N3-X)*C2
410 IF ABS(K1)>ABS(K2) THEN 440
420 LET K2=K1
430 LET X5=X
440 NEXT X
450 IF INT(X5+.0001)=INT(X5+.001) THEN N1=INT(X5+.001) ELSE N1=X5
460 LET N2=N3-N1
470 GO TO 270
480 LET N1=X
490 LET N2=N3-X
500 GO TO 270
510 IF N2=0 THEN 700
520 LET K2=N2*C2*100
530 IF A>C3 THEN 570
540 LET L1=0
550 LET M1=N2
560 GO TO 590
570 LET L1=N2
580 LET M1=(100*N2)
590 FOR X=L1 TO M1 STEP .1
600 IF X*C1+N2*C2=(N2+X)*C3 THEN 670
610 LET K1=(N2+X)*C3-X*C1-N2*C2
620 IF ABS(K1)>ABS(K2) THEN 650
630 LET K2=K1
640 LET X5=X
650 NEXT X
660 GO TO 750
670 LET N1=X
680 LET N3=N1+N2
690 GO TO 270
700 C1= =C2
710 N1= =N2
720 GO TO 520
730 LET N3=N1+N2
740 GO TO 270
750 IF INT(X5+.0001)=INT(X5+.001) THEN N1=INT(X5+.001) ELSE N1=X5
760 LET N3=N2+N1
770 GO TO 270
780 PRINT 'ANYTHING ELSE (YES OR NO)';
790 INPUT A$
800 IF A$='YES ' THEN 100
810 END
```

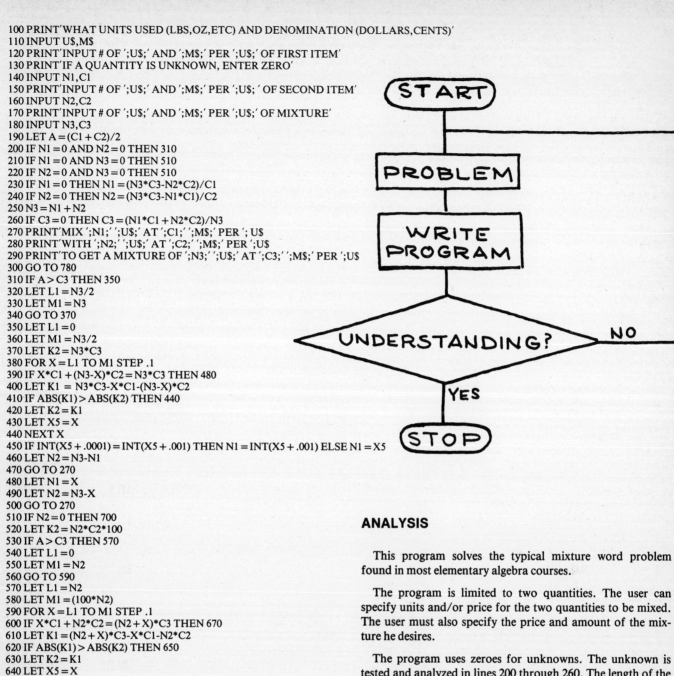

ANALYSIS

This program solves the typical mixture word problem found in most elementary algebra courses.

The program is limited to two quantities. The user can specify units and/or price for the two quantities to be mixed. The user must also specify the price and amount of the mixture he desires.

The program uses zeroes for unknowns. The unknown is tested and analyzed in lines 200 through 260. The length of the program is due to the large number of possible combinations.

The mathematics is elementary and the user is invited to peruse the program, following a line of algorithmic thinking for a specified variable.

Synthetic division is a process for factoring a polynomial by guessing whether a number is a factor. That number is then analyzed in conjunction with the coefficients to produce a remainder and the coefficients of the quotient. Should the remainder be zero the original number would then be a factor. Actually, *root* is a better word here than *factor*.

Write a computer program to perform synthetic division on a given polynomial. Have the computer accept the polynomial's coefficients as well as the guess for a root. Have the computer indicate when the remainder is zero. Have the computer type out the quotient polynomial and the remainder if it is non-zero. Remember the number used in the test and the binomial factor usually differ by a sign.

Be sure to research the process carefully before beginning.

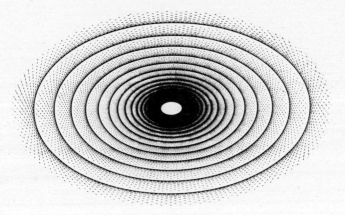

Reference:
M. Keedy, **Algebra and Trigonometry,** p. 177.

```
INPUT THE DEGREE OF THE POLYNOMIAL F(X)
? 3
STARTING WITH X↑ 3 INPUT THE COEFFICIENTS
? 1
? 4
? 8
? 16
INPUT THE DIVISOR IN THE FORM (X+   )? 4
  1 X↑ 3+ 4 X↑ 2+ 8 X↑ 1+ 16 X↑ 0+
QUOTIENT IS  1X↑ 2 +  0X↑ 1 +  8X↑ 0 +
REMAINDER IS -16
(X +  4) IS NOT A FACTOR!


ANY MORE TO DO
? YES
INPUT THE DEGREE OF THE POLYNOMIAL F(X)
? 2
STARTING WITH X↑ 2 INPUT THE COEFFICIENTS
? 1
? 7
? 12
INPUT THE DIVISOR IN THE FORM (X+   )? 4
  1 X↑ 2+ 7 X↑ 1+ 12 X↑ 0+
QUOTIENT IS  1X↑ 1 +  3X↑ 0 +
(X+  4) IS A FACTOR!


ANY MORE TO DO
? NOPE
```

```
100 DIM A(15),B(15)
110 PRINT 'INPUT THE DEGREE OF THE POLYNOMIAL F(X)'
120 INPUT N
130 LET W = N-1
140 PRINT 'STARTING WITH X↑';N;' INPUT THE COEFFICIENTS'
150 FOR X = 1 TO N+1
160 INPUT A(X)
170 NEXT X
180 PRINT 'INPUT THE DIVISOR IN THE FORM (X+  )';
190 INPUT D
200 FOR X = 1 TO N+1
210 PRINT A(X);' X↑';W+1;' +';
220 LET W = W-1
230 NEXT X
240 LET D = -D
250 PRINT
260 PRINT 'QUOTIENT IS ';
270 LET B(1) = A(1)
280 LET W = N-1
290 FOR X = 1 TO N
300 LET T = X+1
310 LET B(T) = A(T) + (D*B(X))
320 PRINT B(X);' X↑';W;' + ';
330 LET W = W-1
340 NEXT X
350 PRINT
360 IF B(N+1) = 0 THEN 390
370 PRINT 'REMAINDER IS ';B(N+1)
375 PRINT '(X + ';-D;') IS NOT A FACTOR!'
380 PRINT
385 GO TO 400
390 PRINT '(X + ';-D;') IS A FACTOR!'
400 PRINT
410 PRINT 'ANY MORE TO DO'
420 INPUT A$
430 IF A$ = 'YES' THEN 100
440 END
```

ANALYSIS

A close look at synthetic division makes it obvious that it is not a difficult arithmetic procedure. It is, however, tedious and time consuming.

The program uses a rather inefficient input loop. The coefficients could just as easily have been inputed from a data statement. An attempt was made to make the program interactive, so that roots could be searched out.

The variable W keeps track of the exponents. The 200 loop ending in 230 simply prints out the polynomial. Line 240 changes the sign of the binomial factor form so that input may be made in the form $(x + D)$.

The actual division is done in line 310. A subscripted variable is used to store the elements until we are ready to print out the quotient. The remainder is tested in line 360 and printed in the next line. If the remainder is zero that fact is indirectly announced by line 390. Notice that the value of D is again changed by a sign to get the output to be consistent with the binomial form which we inputed.

19   Systems Of Equations Up To Four Unknowns

Write a program to solve a system of equations with up to four unknowns. You may use any algorithm you like. Cramer's Rule with the use of determinants is one way. Gaussian elimination is a second. A trial and error axiom is to be discouraged.

There are some iterative algorithms available, but they may not be as easy as Cramer's Rule.

The use of matrices is also a possibility. For example, if

$$ax + by + cz = d$$
$$ex + fy + gz = h$$
$$ix + jy + kz = l$$

then in matrix notation

$$\begin{pmatrix} a & b & c \\ e & f & g \\ i & j & k \end{pmatrix} * \begin{pmatrix} x \\ y \\ z \end{pmatrix} = \begin{pmatrix} d \\ h \\ l \end{pmatrix}$$

which implies that:

$$A \quad * \quad X \quad = \quad K$$

$$A^{-1} * A * X = A^{-1} * K$$
$$1 * X = A^{-1} * K$$
$$X = A^{-1} * K$$

With the use of matrix instructions or an algorithm for finding an inverse, the problem is quite simple.

Note: Be sure to test the system for consistency and existence of a solution!

References:

Computer Oriented Mathematics by NCTM, Chapter 3, page 95.

Mathematics for High School: Introduction to Matrix Algebra, SMSG.

```
 .99999982      -2.9999998      1.9999999
-2.9999999       2.9999999      -.99999996
 2.0            -.99999999       2.4835265E-09
SOLUTIONS ARE

 .99999981
-1.9999999
 2.0
```

```
15 DIM A(3,3),X(3,1),C(3,1),B(3,3)
25 MAT READ A,X
30 MAT B = INV(A)
45 MAT PRINT B
50 MAT C = B*X
60 PRINT 'SOLUTIONS ARE '
70 MAT PRINT C
80 DATA 1,2,3,2,4,5,3,5,6,3,4,5
100 END
```

ANALYSIS

The program is elementary with the MATRIX statements in BASIC. Line 25 reads the 3×3 matrix A and the column vector X. Line 30 inverts A and stores the inverse in a matrix B. Line 45 prints the inverse. Notice the truncation which takes place. The program could be modified to clean up the printout. This is left as an exercise.

Line 50 sets up the solution by multiplying the inverse by the column vector. The theory is fully explained on the student problem sheet. Notice the word MAT is necessary in line 50 to effect matrix multiplication.

Line 70 prints the answer in column vector form.

A rope over the top of a fence has the same length on each side. The rope weighs one-third pound per foot. On one end hangs a monkey holding a banana, and on the other end is a weight equal to the weight of the monkey. The banana weighs two ounces per inch.

The rope is as long as the age of the monkey. The weight of the monkey in ounces is as much as the age of the monkey's mother. The combined ages of the monkey and the monkey's mother are 30 years. Half the weight of the monkey, plus the weight of the banana, is a fourth as much as the weight of the weight and the weight of the rope.

The monkey's mother is half as old as the monkey will be when it is three times as old as its mother was when she was half as old as the monkey will be when it's as old as its mother will be when she is four times as old as the monkey was when it was twice as old as its mother was when she was a third as old as the monkey was when it was as old as its mother was when she was 30 times as old as the monkey was when it was a fourth as old as it is now.

How long is the banana? There is a solution possible with the information given above.

Write a program to compute how many permutations or combinations there are for *N* things taken *R* at a time.

The user should be able to specify which measure he wants. The computer should then simply print out how many. This is a simple matter of learning some formulas. These formulas contain factorials and these must be computed prior to inclusion in the final computation. The factorial computation should be done in a loop prior to the actual plugging-in to the formula.

References:

Fred Mosteller, **Probability and Statistics**, p. 19–47

Henry Alder, **Introduction to Probability and Statistics**, pp. 60–7

```
DO YOU WISH PERMUTATIONS OR COMBINATIONS (P OR C)
? P
PERMUTATIONS - HOW MANY OBJECTS AND HOW MANY AT A TIME
? 15,4
THERE ARE   32760.0 WAYS TO PERMUTE
 15 THINGS   4 AT A TIME!

DO YOU WISH PERMUTATIONS OR COMBINATIONS (P OR C)
? C
COMBINATIONS - HOW MANY OBJECTS AND HOW MANY AT A TIME
? 12,6
THERE ARE  924 WAYS TO COMBINE
 12 THINGS TAKEN  6 AT A TIME!!

DO YOU WISH PERMUTATIONS OR COMBINATIONS (P OR C)
? STOP
 PROGRAM STOPPED.
```

```
100 PRINT
110 PRINT 'DO YOU WISH PERMUTATIONS OR COMBINATIONS (P OR C)'
120 INPUT A$
130 IF A$='C' THEN 160
140 PRINT 'PERMUTATIONS - HOW MANY OBJECTS AND HOW MANY AT A TIME'
150 GO TO 170
160 PRINT 'COMBINATIONS - HOW MANY OBJECTS AND HOW MANY AT A TIME'
170 INPUT N,R
180 LET F=G=H=1
190 LET B=N-R
200 FOR X=1 TO N
210 LET F=F*X
220 NEXT X
230 FOR Y=1 TO B
240 LET G=G*Y
250 NEXT Y
260 IF A$='C' THEN 300
270 PRINT 'THERE ARE ';F/G;' WAYS TO PERMUTE '
280 PRINT N;' THINGS ';R;' AT A TIME!'
290 GO TO 100
300 FOR Z=1 TO R
310 LET H=H*Z
320 NEXT Z
330 PRINT 'THERE ARE ';F/(H*G);' WAYS TO COMBINE '
340 PRINT N;' THINGS TAKEN ';R; ' AT A TIME!!'
350 GO TO 100
360 END
```

ANALYSIS

This program uses the old familiar formulas for permutations and combinations. The loops beginning at 200 and 230 are used to compute the factorials.

Notice how the partial product is reset to 1 in line 180 for each run.

You've all had the opportunity to make use of log tables, during your mathematical lives. Ever wonder where those tables come from? Here's a chance for you to make a set for yourself.

Limit the table to the integers between 1 and 100. Of course you may not use the built-in LOG function available in BASIC. However, with the knowledge that:

$$\log_{10} x = a \text{ only when } 10^a = x$$

one should easily be able to proceed by solution of that relation.

Since:

$$\log_{10} 100 = 2$$

it only stands to reason that:

$$10^2 = 100$$

Try to get the output in the following form (it's concise):

1	0	410	788	1136	1458	1758	2038	2301	2550	2785
2	3009	3221	3423	3616	3801	3978	4149	4313	4471	4623
3	4770	4913	5051	5184	5314	5440	5562	5681	5797	5910
4	6020	6127	6232	6334	6434	6532	6627	6721	6812	6902
5	6989	7075	7160	7242	7324	7403	7482	7558	7634	7708
6	7781	7853	7924	7993	8062	8129	8195	8261	8325	8388
7	8451	8512	8573	8633	8692	8751	8808	8865	8921	8976
8	9031	9085	9138	9191	9243	9294	9345	9395	9445	9494
9	9542	9590	9638	9685	9731	9777	9823	9868	9912	9956

IMPORTANT: Omit the characteristic. You may retain the decimal point if you wish. FOUR places please.

```
110 LET W=0
120 FOR X=1 TO 9
130 PRINT X;
140 FOR Y=0 TO 9
150 LET Z=(10*X+Y)*.1
160 IF ABS(10↑W-Z)>.001 THEN 210
170 LET T=INT(10000*W)
190 PRINT TAB(6*Y+4);T;
200 GOTO 230
210 LET W=W+.0001
220 GOTO 160
230 NEXT Y
240 PRINT
250 NEXT X
260 END
READY
```

ANALYSIS

The program works by testing decimals as powers of 10. When they produce the desired number they are printed.

Line 160 tests the power. Line 170 takes only the first four places of that power. Line 180 prints and orders the format of the columns.

References:

Samuel Selby, **Standard Mathematical Tables,** pp. 1–6.

J.S. Meyer, **Fun With Mathematics,** pp. 90–99.

W.L. Bashaw, **Mathematics for Statistics,** pp. 253–264.

Listing of Permutations and Combinations

There is sometimes a need to know more than just the number of combinations or permutations of a given number of elements. It is sometimes necessary to list the permutations or combinations that start with a given letter or series of letters.

Write a program to type out all possible permutations or combinations that have a certain property. Perhaps a sequence of items can be specified in a DATA statement. The program should also include the total possible permutations or combinations from the standard formula.

References:

Fred Mosteller, **Probability and Statistics,** pp. 19–47.

Henry Alder, **Introduction to Probability and Statistics,** pp. 60–7.

1	2	3	4		1	2	4	3		1	3	2	4		1	3	4	2		1	4	2	3
1	4	3	2		2	1	3	4		2	1	4	3		2	3	1	4		2	3	4	1
2	4	1	3		2	4	3	1		3	1	2	4		3	1	4	2		3	2	1	4
3	2	4	1		3	4	1	2		3	4	2	1		4	1	2	3		4	1	3	2
4	2	1	3		4	2	3	1		4	3	1	2		4	3	2	1					

```
10 FOR A = 1 TO 4
20 FOR B = 1 TO 4
30 FOR C = 1 TO 4
40 FOR D = 1 TO 4
50 IF A = B OR A = C OR A = D THEN 200
60 IF B = C OR B = D OR C = D THEN 200
70 PRINT A;B;C;D; '   ';
200 NEXT D
210 NEXT C
220 NEXT B
230 NEXT A
300 END
```

ANALYSIS

It's a sneaky way to do it but it works. The algorithm simply takes all numbers from 1,111 to 4,444 and throws away any that have a matching digit. This is done by statements 50 through 62.

It can be expanded to do 5 or 6 numbers permuted by changing both the end limit of the loops and the comparison test in lines 50 and 60.

Write a program to accept the equation of any quadratic. Test the equation to see if it will produce a parabola. Reject the equation if it will not be a parabola.

Have the computer type out the zeroes of the curve; the high point or low point; the equation of the axis of symmetry; whether it is concave upward or downward; along with the sum and product of the roots. Have the computer plot the graph of the parabola indicating the zeroes of function with symbols other than the ones you're plotting with. Have the computer include the axis of symmetry.

Be sure the roots are real before attempting solution on the real number axes. You may use the quadratic formula for the zeroes or you may wish to use a program referred to in another problem. (See **Zeroes of a Function by Iteration**).

Do some preliminary research in an algebra text on conic sections. Save yourself some time by reducing the problem to its essentials.

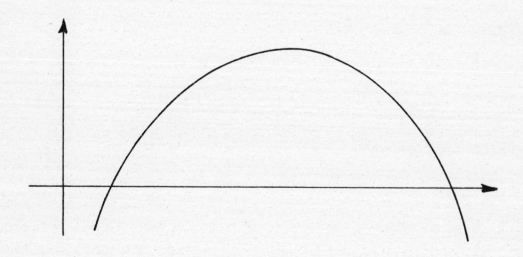

References:

W.A. Wilson, **Analytic Geometry,** pp. 98–108.

```
INPUT CØEFFICIENTS ØF QUADRATIC IN FØRM AX↑2+BX+C=0
? 2,6,8
NØ REAL RØØTS
ANYMØRE TØ DØ (YES ØR NØ)
? YES
INPUT CØEFFICIENTS ØF QUADRATIC IN FØRM AX↑2+BX+C=0
? 1,2,-8
CØNCAVE UP
RØØTS ØF QUADRATIC ARE   2  AND  -4
SUM ØF RØØTS IS  -2
PRØDUCT ØF RØØTS IS  -8
AXIS ØF SYMMETRY IS X = -1
HIGH ØR LØW PØINT IS (-1 , -9)
                        GRAPH ØF PARABØLA
                   (*NØTE*X-AXIS IS VERTICAL)
SCALE ØN Y-AXIS IS 1 SPACE =  1 SPACE
          ++++++++++++++++++++++++++++++++++++++++++++++++++++++++++++++
 -8       +            +                                      *
 -7       +            +                              *
 -6       +            +                      *
 -5       +            +        *
 -4       +            X
 -3       +        *   +
 -2       +     *      +
 -1       +  *         +
  0       +     *      +
  1       +        *   +
  2       +            X
  3       +            +        *
  4       +            +                *
  5       +            +                      *
  6       +            +                              *
ANYMØRE TØ DØ (YES ØR NØ)
? NØ
```

30 PRINT´INPUT COEFFICIENTS OF QUADRATIC IN FORM AX↑2+BX+C=0´
40 INPUTA,B,C
50 IF A=0 THEN 440
60 D=B↑2-4*A*C
70 IF D<0 THEN 460
80 X1=(-B+SQR(D))/(2*A)
90 X2=(-B-SQR(D))/(2*A)

```
110 IF A<0 THEN PRINT'CONCAVE  DOWN' ELSE PRINT'CONCAVE UP'
120 PRINT 'ROOTS OF QUADRATIC ARE ';X1;' AND 'X2
130 PRINT 'SUM OF ROOTS IS ';-B/A
140 PRINT 'PRODUCT OF ROOTS IS ';C/A
150 E = -B/(2*A)
160 PRINT 'AXIS OF SYMMETRY IS X = ';E
170 X3 = A*E↑2 + B*E + C
180 PRINT 'HIGH OR LOW POINT IS (';E;' , ';X3;')'
200 PRINT TAB(25);'GRAPH OF PARABOLA'
210 PRINT TAB(21);'(*NOTE*X-AXIS IS VERTICAL)'
220 LET X = E-7
230 Y = A*X↑2+B*X+C
233 B$='+'
234 C$='*'
240 IF ABS(Y) + ABS(X3)>65 THEN 400 ELSE K=1
245 PRINT'SCALE ON Y-AXIS IS 1 SPACE = '; 1/K;' SPACE'
246 FOR Q=1 TO 63
247 PRINT TAB(8);'+';
248 NEXT Q
250 IF A<0 AND X3<=0 THEN 330
260 IF A>0 AND X3>=0 THEN 480
270 IF A<0 AND X3>=0 THEN 540 ELSE 670
280 IF B1=B2 THEN 800
285 IF INT(B1)=INT(B2) THEN C$=' '
290 PRINT X;TAB(8);TAB(B1);B$;TAB(B2);C$
300 X = X + 1
310 Y = A*X↑2+B*X+C
314 B$='+'
317 C$='*'
320 RETURN
330 W=Y*K
340 B1=ABS(W-K*Y)+8
350 B2=ABS(W) +8
380 GOSUB 280
390 IF X<=E+7 THEN 340 ELSE 830
400 M = ABS(Y) + ABS(X3)
410 K=1/(INT(M/65+1))
430 GO TO 245
440 PRINT 'LINEAR EQUATION-NOT QUADRATIC'
450 GO TO 830
460 PRINT 'NO REAL ROOTS'
470 GO TO 830
480 B1=8
490 B2=(K*Y)+8
500 B$==C$
520 GOSUB 280
530 IF X<=E+7 THEN 490 ELSE 830
540 W=Y*K
550 B1=K*Y+ABS(W)+8
560 B2=ABS(W)+8
570 B$==C$
590 IF Y>0 THEN 630
600 GOSUB 280
610 IF X<=E+7 THEN 550 ELSE 830
630 B1==B2
635 B$==C$
640 GO TO 600
670 W=X3*K
680 B1= K*Y+ABS(W)+8
690 B2=ABS(W) +8
700 B$==C$
710 IF Y>0 THEN 730
720 GOSUB 280
725 IF X<=E+7 THEN 680 ELSE 830
730 B1==B2
733 B$==C$
737 GO TO 720
800 B$='X'
810 C$=' '
820 GO TO 290
830 PRINT 'ANYMORE TO DO (YES OR NO)'
840 INPUT A$
850 IF A$='YES' THEN 30
900 END
```

ANALYSIS

This program computes the roots, sum and product thereof, axis of symmetry and turning point for any quadratic equation. It also graphs the resulting parabola.

The quadratic formula and the resulting formulas for product and sum of roots are used in lines 60 through 180. Line 50 tests for a second degree equation.

The rest of the program is taken up with drawing the graph. Line 240 scales down the graph so it will fit on a single sheet of TTY paper. Lines 250 and 270 determine how the parabola is situated with respect to the x-axis. Care is taken to note where the graph crosses the x-axis; when this happens the * takes precedence over the + in that case.

The location of the x-axis relative to the left margin is determined by the variables *B1* and *B2* to insure the printing out of values below the x-axis.

25 Solutions to the Cubic Equation

Write a program to find all the roots of a cubic equation of the form:

$$ax^3 + bx^2 + cx + d = 0$$

where a, b, c and d are inputed. Remember all the roots need not be real. If only one root is real then the other two will be complex.

Reduce the cubic to a quadratic and then use the quadratic formula to find the complex roots. Be sure to express them in the form:

$$a + bi$$

Devise a test to determine the number of real roots.

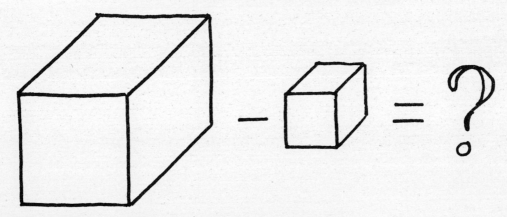

```
INPUT COEFFICIENTS OF AX↑3+BX↑2+CX+D=0
? 1,-8,4,17
INPUT LOWER & UPPER LIMITS & INTERVAL WIDTH TO INVESTIGATE
? -10,10,1
X-VALUE          SIGN
   -10            -
   -9             -
   -8             -
   -7             -
   -6             -
   -5             -
   -4             -
   -3             -
   -2             -
   -1             +
    0             +
    1             +
    2             +
    3             -
    4             -
    5             -
    6             -
    7             -
    8             +
    9             +
   10             +
```

References:

Kemeny and Kurtz, **Basic Programming,** pp. 50–53.

Be sure to review what you've learned about cubics in an advanced secondary math text. It'll save programming time.

```
NO EXACT ROOTS FOUND**THE APPROXIMATE ROOTS FOLLOW
THE ROOTS ARE  7.099255 ,  2.0620340 AND -1.1612891
ANYTHING ELSE (YES OR NO)? NO
```

```
100 PRINT'INPUT COEFFICIENTS OF AX↑3+BX↑2+CX+D=0'
110 INPUT A,B,C,D
120 PRINT 'INPUT LOWER & UPPER LIMITS & INTERVAL WIDTH TO INVESTIGATE'
130 LET F=F2=K2=K3=0
140 INPUT L1,H1,W
150 IF H1<L1 THEN L1==H1
160 LET K1=A*L1↑3+B*L1↑2+C*L1+D
170 PRINT 'X-VALUE','SIGN'
180 FOR X=L1 TO H1 STEP W
190 LET M1=A*X↑3+B*X↑2+C*X+D
200 LET K3=K3+1
210 IF SGN(M1)=0 THEN 280
220 IF SGN(M1)=1 THEN A$='+' ELSE A$='-'
230 PRINT TAB(3);X;TAB(16);A$
240 IF F=0 AND SGN(M1)><SGN(K1) THEN 330
250 LET K2=K2+1
260 LET K1=M1
270 GO TO 490
280 LET A$='A ROOT'
290 LET K3=K3+1
300 R3=X
310 F=F+1
320 GO TO 230
330 IF F2><0 THEN 490
340 LET L=X-W
350 LET H=X
360 LET F1=0
370 LET X1=(L+H)/2
380 LET M2=A*X1↑3+B*X1↑2+C*X1+D
390 IF SGN(M2)=0 THEN 440
400 LET M3=A*L↑3+B*L↑2+C*L+D
410 IF SGN(M2)=SGN(M3) THEN L=X1 ELSE H=X1
420 LET F1=F1+1
430 IF F1>50 THEN 470 ELSE 370
440 LET R3=X1
450 LET F=F+1
460 GO TO 490
470 LET S1=X1
480 LET F2=F2+1
490 NEXT X
500 IF K2=K3 THEN 600
510 IF F=0 THEN 670
520 LET B1=A*R3+B
530 LET C1=B1*R3+C
540 LET D1=B1↑2-4*A*C1
550 IF D1<0 THEN 620
560 LET R1=(-B1+SQR(D1))/(2*A)
570 LET R2=(-B1-SQR(D1))/(2*A)
580 PRINT'THE ROOTS ARE ';R1;' , ';R2;' AND ';R3
590 GO TO 700
600 PRINT'NO ROOTS FOUND IN INTERVAL-SELECT NEW LIMITS'
610 GO TO 120
620 PRINT'THERE IS ONE REAL ROOT ** ';R3
630 PRINT'IN COMPLEX FORM, THE OTHER TWO ROOTS ARE **'
640 PRINT -B1/(2*A);'+';SQR(ABS(D1))/(2*A);' I';
650 PRINT' AND ';-B1/(2*A);'-';SQR(ABS(D1))/(2*A);' I'
660 GO TO 700
670 R3=S1
680 PRINT'NO EXACT ROOTS FOUND**THE APPROXIMATE ROOTS FOLLOW'
690 GO TO 520
700 PRINT 'ANYTHING ELSE (YES OR NO)';
710 INPUT A$
720 IF A$='YES' THEN 100
730 END
```

ANALYSIS

This program finds the three roots of any third-degree (cubic) equation. It prints roots whether they are real or complex. Complex roots are printed in rectangular form.

The user initiates the search for the roots by specifying an interval to search. Starting with the lower limit the computer prints the signs so that the user can see the changes in sign and locate the approximate location of the roots.

If a zero is found for the values actually printed this is so indicated. If no sign change was encountered the user is asked to select new limits.

If an exact root is found this is used in lines 520 and 530 to reduce the cubic to a quadratic by synthetic division. This resulting quadratic is solved using the quadratic formula in lines 540 and 570.

If a change is found in the sign but no exact root encountered, the root is approximated by repeated halving of the interval in which the root is known to lie. This is done in lines 370 through 430. This approximate root is then used in the synthetic division.

Using the figure below find the values of X and Y. Write a program to solve the problem. The program could involve an algorithm for synthetic division (see problem with that name) or a simple cranking out of *Pythagorean Triples*. (See problem of same name).

A lot if preliminary work is needed here, so be sure to reduce the problem to its programmable state. All the information necessary for the solution is given in the diagram, believe it or not.

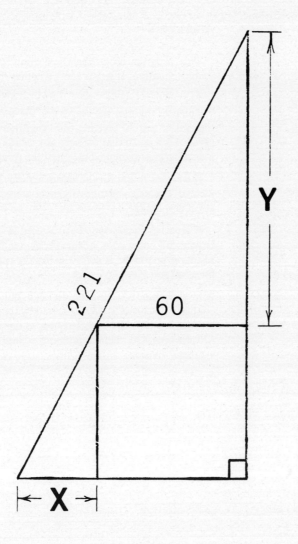

SOLUTION TO RIGHT TRIANGLE AND THE SQUARE

```
X =  25 & Y = 144  - SIDES ARE: BASE =  85 & ALTITUDE = 204
X =  44 & Y = 135  - SIDES ARE: BASE = 104 & ALTITUDE = 195
X =  80 & Y = 111  - SIDES ARE: BASE = 140 & ALTITUDE = 171
X = 111 & Y =  80  - SIDES ARE: BASE = 171 & ALTITUDE = 140
X = 135 & Y =  44  - SIDES ARE: BASE = 195 & ALTITUDE = 104
X = 144 & Y =  25  - SIDES ARE: BASE = 204 & ALTITUDE =  85
```

```
10 PRINT "SOLUTION TO RIGHT TRIANGLE AND THE SQUARE"
20 PRINT
25 Z=221*221
30 FOR X=0 TO 160
40 FOR Y=0 TO 160
50 IF (X+60)↑2+(Y+60)↑2<>Z THEN 70
60 PRINT "X =";X;"& Y =";Y;
65 PRINT " - SIDES ARE: BASE =";60+X;"& ALTITUDE =";60+Y
70 NEXT Y
80 NEXT X
90 END
```

ANALYSIS

This problem can be solved easily by using a brute search procedure or by using the solution of a quartic equation using synthetic division. Careful use of the diagram is necessary.

From the diagram it can be seen that

$$(X + 60)^2 + (Y + 60)^2 = 221^2 \qquad (1)$$

The brute search procedure simply tries integer values of X and Y such that each side is varied over a more than adequate range. If X and Y are each tried from 1 to 180, then each side will vary from 60 to 220, more than adequate. Any values of X and Y that combine to satisfy the equation present a possible solution.

A more elegant solution can be obtained by noting that

$$\frac{Y}{60} = \frac{60}{X}$$

giving

$$X \cdot Y = 3600 \qquad (2)$$

If (1) and (2) are combined to yield an equation in X, the following quartic is obtained:

$$x^4 + 120x^3 - 41,641x^2 + 432,000x - 12,960,000 = 0 \quad (3)$$

This equation can be solved using the synthetic division program given elsewhere in this book.

The reader is invited to develop a test that will stop execution before the *mirror image* solutions are computed. Such a test will save considerable computer time.

Pythagorean Triples

Every student knows the pythagorean theorem and its importance in working with right triangles. In fact there is hardly a student who isn't familiar with the numbers (3,4,5) or (5,12,13). Some students know others. Few students can give more than a couple of these *Pythagorean Triples*. Obviously a *Pythagorean Triple* is a set of numbers which satisfy the relationship:

$$x^2 + y^2 = z^2.$$

Write a program to list as many triples of this sort as you deem feasible. A simple way would be to test all triples and PRINT only those which satisfy the theorem. This is an extremely inefficient algorithm. The computer could test as many as a million numbers before it encountered a *Pythagorean Triple*.

The student is invited to find what are known as *generators*, that is, algebraic expressions which when fulfilled supply the desired triples every time. One useful *generator* could be any two natural numbers u and v (one greater than the other), such that the sum of their squares produces one element of the triple; the difference between their squares produces a second; and twice their product produces the last. The natural numbers may differ by any number of units and it is intriguing to note the relationships which *pop up*.

After you've listed the triples inspect them closely. You'll find many interesting relationships. Did you know that every set of triples has at least one element that is divisible by 3, one that is divisible by 4 and one that is divisible by 5?

References:

Ivan Niven, **Introduction to the Theory of Numbers,** pp. 1-9.

O. Ore, **Invitation to Number Theory.**

B.M. Stewart, **Theory of Numbers,** pp. 153-6.

GENERATION OF PYTHAGOREAN TRIPLES!!!

TRIPLES			MULT OF	U AND V
3	4	5		(1, 2)
6	8	10	(3, 4, 5)	(1, 3)
8	15	17		(1, 4)
10	24	26	(5, 12, 13)	(1, 5)
12	35	37		(1, 6)
5	12	13		(2, 3)
12	16	20	(3, 4, 5)	(2, 4)
20	21	29		(2, 5)
24	32	40	(3, 4, 5)	(2, 6)
7	24	25		(3, 4)
16	30	34	(8, 15, 17)	(3, 5)
27	36	45	(3, 4, 5)	(3, 6)
9	40	41		(4, 5)
20	48	52	(5, 12, 13)	(4, 6)
11	60	61		(5, 6)

```
010 PRINT 'GENERATION OF PYTHAGOREAN TRIPLES!!!'
20 PRINT
30 PRINT 'TRIPLES',' ',' ','MULT OF','U AND V'
50 PRINT
60 FOR U = 1 TO 10
70 FOR V = U + 1 TO 6
80 LET A = 2*U*V
90 LET B = U↑2 + V↑2
100 LET C = V↑2-U↑2
110 FOR N = 10 TO 2 STEP -1
120 IF A/N < > INT(A/N) THEN 240
130 IF B/N < > INT(B/N) THEN 240
140 IF C/N < > INT(C/N) THEN 240
150 LET F = A/N
160 LET G = B/N
170 LET H = C/N
180 LET K = MAX(A,C)
190 LET J = MIN(C,A)
200 LET S = MAX(F,H)
210 LET T = MIN(H,F)
220 PRINT J,K,B,'(';T;',';S;',';G;')','(';U;',';V;')'
230 GO TO 270
240 NEXT N
250 LET K = MAX(A,C)
260 LET J = MIN(C,A)
265 PRINT J,K,B,' ','(';U;',';V;')'
270 NEXT V
280 NEXT U
300 END
```

ANALYSIS

The famous triple $2uv$, $u^2 - v^2$ and $u^2 + v^2$ is used to generate the Pythagorean numbers. This triplet will work whenever $u > v$.

The program generates the triples in lines 80 through 100. The N loop beginning in 110 tests for the triple being a multiple of one already printed. If it is original it prints out the triple along with the values of u and v. If it is a multiple it prints out the triple it is a multiple of.

Notice that every set of Pythagorean triples has elements which are divisible by 3, 4, and 5. For example 9, 40, 41 has 9 which is divisible by 3 and 40 which is divisible by 4 and 5.

Notice the use of the MAX and MIN functions. They return the larger or smaller of the arguments. Although the functions are not available in all forms of BASIC, routines can be written to simulate them.

Notice that the product of all three numbers is always a multiple of 60.

28 Area of a Triangle by a Coordinate Geometry Method

Write a program to compute the area of a triangle. Assume that the polygon has a known but variable number of sides.

As input you will need the number of sides of the polygon along with the coordinates of its vertices. It will not necessarily be a regular polygon, nor will it be convex. Using only the coordinates of each vertex and the number of sides you must compute the area.

This will require an algorithm which is already in existence. The development of the algorithm for the area is not unusually difficult and can be done by you. However, you can probably find it in all but the most elementary texts on plane or coordinate geometry.

Why not test your skill at programming graphics output? The student can extend the scope of the problem by computing the area of a polygon using more than three sides.

References:

Edwin Moise, **Geometry,** pp. 371–406.

W.A. Wilson, **Analytical Geometry,** pp. 8–27.

Hulian Mancill, **Modern Analytical Trigonometry,** pp. 27–56.

```
THIS PROGRAM WILL ACCEPT THE COORDINATES OF THE VERTICES
OF A TRIANGLE. IT WILL THEN COMPUTE THE AREA AND
THE TYPE OF TRIANGLE IN QUESTION!!!

INPUT THE COORDINATES OF THE VERTICES
? 0,0,3,0,0,3
THE AREA IS  4.5

IT'S A RIGHT TRIANGLE
THE TRIANGLE IS ISOSCELES!!
ANGLES IN DEGREES ARE  45  45  90

READY

RUNH
THIS PROGRAM WILL ACCEPT THE COORDINATES OF THE VERTICES
OF A TRIANGLE. IT WILL THEN COMPUTE THE AREA AND
THE TYPE OF TRIANGLE IN QUESTION!!!

INPUT THE COORDINATES OF THE VERTICES
? 0,1,4,5,-4,6
THE AREA IS  18

THE TRIANGLE IS SCALENE!
ANGLES IN DEGREES ARE  44.2152  52.125  83.6598
```

```
10 PRINT 'THIS PROGRAM WILL ACCEPT THE COORDINATES OF THE VERTICES'
20 PRINT 'OF A TRIANGLE. IT WILL THEN COMPUTE THE AREA AND'
30 PRINT 'THE TYPE OF TRIANGLE IN QUESTION!!!'
35 PRINT
40 PRINT 'INPUT THE COORDINATES OF THE VERTICES'
45 INPUT X1,Y1,X2,Y2,X3,Y3
50 LET S1 = SQR((Y2-Y1)↑2 + (X2-X1)↑2)
55 LET S2 = SQR((Y3-Y1)↑2 + (X3-X1)↑2)
57 LET S3 = SQR((Y3-Y2)↑2 + (X3-X2)↑2)
60 LET S = (S1 + S2 + S3)/2
65 LET T = SQR(S*(S-S1)*(S-S2)*(S-S3))
70 PRINT 'THE AREA IS ';T
75 PRINT
77 IF S1↑2 + S2↑2 = S3↑2 THEN PRINT 'ITS A RIGHT TRIANGLE!'
78 IF S2↑2 + S3↑2 = S1↑2 THEN PRINT 'ITR A RIGHT TRIANGLE!'
79 IF S1↑2 + S3↑2 = S2↑2 THEN PRINT 'ITS A RIGHT TRIANGLE!'
80 LET P = 0
81 IF S1 < > S2 THEN 85
82 LET P = P + 1
85 IF S2 < > S3 THEN 105
90 LET P = P + 1
105 ON P + 1 GO TO 200,210,220
150 GO TO 300
200 PRINT 'THE TRIANGLE IS SCALENE!'
205 GO TO 300
210 PRINT 'THE TRIANGLE IS ISOSCELES!!'
215 GO TO 300
220 PRINT 'THE TRIANGLE IS EQUILATERAL!!!'
300 LET R = SQR((S-S1)*(S-S2)*(S-S3))/SQR(S)
310 LET A1 = (ATN(R/(S-S1))*2)*57.295779
320 LET A2 = (ATN(R/(S-S2))*2)*57.295779
330 LET A3 = (ATN(R/(S-S3))*2)*57.295779
340 PRINT 'ANGLES IN DEGREES ARE ';A1;A2;A3
400 END
```

ANALYSIS

The program is an example of a lot of tedious yet simple calculation done quickly. The coordinates of the vertices of a triangle are converted into the lengths of the sides in lines 50 through 57.

Line 60 computes the semi-perimeter. The area is computed by Hero's formula which is given below.

$$A = \sqrt{s \bullet (s-a) \bullet (s-b) \bullet (s-c)}$$

where s is the semi-perimeter and a, b, and c are the sides of the triangle.

The lengths of the sides are then compared to one another to establish the type of triangle. Lines 77 through 79 check the sides against the Pythagorean Theorem.

The angles are computed using the formula given below.

$$\tan \frac{\theta}{2} = \sqrt{\frac{(s-b) \bullet (s-c)}{s \bullet (s-a)}} = \frac{r}{(s-a)}$$

where

$$r = \sqrt{\frac{(s-a) \bullet (s-b) \bullet (s-c)}{s}}$$

The formula can be found on page 509 of the Chemical—Rubber Company's *Standard Mathematical Tables*.

Archimedian—Geometric Determination of π

The determination of π has fascinated mathematicians since time began.

One of the more interesting ways to generate it is that used by Archimedes. Although he did it arithmetically, as well as by other methods, this is by far his most famous.

Essentially what he did was compute the perimeters of regular polygons that were inscribed in and circumscribed about a circle. For ease of computation let *your* circle have a radius of one—a unit circle. The circumference will now be 2π.

Start with a square which you circumscribe about the circle. The diagonal of the inner square will equal the side of the outer square; and both will always be one, while the apothem of the outer polygon will always be one.

You must develop an expression for the perimeter of each polygon which involves the apothem and the radius and the number of sides.

Write a program to compute the perimeters of these polygons as their sides are successively doubled in number. It is plain that the outer perimeter will always exceed the circumference of the circle while the inscribed perimeter will always fall a bit short. The ratio of the circumference to the diameter will always contain some form of π. Compute the ratio of the perimeter to the diameter for each polygon and PRINT it. You will find that this ratio will approach π from the left and the right.

Try to avoid an algorithm which contains π. Some trigonometric expression would be desirable.

References:

A.R. Amir-Moez, **More Chips from the Mathematical Log,** p. 75

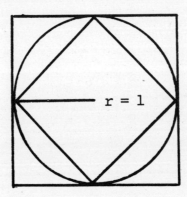

ARCHIMEDEAN DETERMINATION OF PI !

NO. OF SIDES	INSCR PER	CIRCUM PER
4	2.8284272	4.0000001
8	3.0614675	3.3137086
16	3.1214452	3.1825979
32	3.1365485	3.1517249
64	3.1403312	3.1441184
128	3.1412773	3.1422237
256	3.1415138	3.1417504
512	3.1415729	3.1416321
1024	3.1415877	3.1415927
2048	3.1415914	3.1415927
4096	3.1415923	3.1415927
8192	3.1415925	3.1415927
16384	3.1415925	3.1415927
32768	3.1415927	3.1415927

```
50 PRINT 'ARCHIMEDEAN DETERMINATION OF PI!'
60 PRINT
70 PRINT 'NO. OF SIDES','INSCR PER','CIRCUM PER'
80 PRINT
100 FOR X=2 TO 15
105 LET N=2↑X
110 LET D=360/N
120 LET T=3.1415927*(D/180)
130 LET A=2*N*SIN(T/2)
140 LET B=2*N*TAN(T/2)
150 PRINT N,A/2,B/2
155 IF A=B THEN 170
160 NEXT X
170 END
```

ANALYSIS

The limit on the perimeter here is 2^{15} sides.

Line 110 computes the central angle subtended by a side of the polygon. Line 120 converts the angle to radians. It bothers some people to see a decimal approximation to π appearing in a program to generate π. Perhaps it would be better to convert to radians by one of the many methods available so that students wouldn't be alarmed at the technique. An isosceles triangle is formed by the radii. The altitude to the triangle thus formed has hypotenuse 1 and a leg equal to half the side of the edge of the original polygon. In that triangle the following is true:

$$\sin \theta = s/2$$

where s is the side of the inscribed polygon and θ is half the central angle D.

These facts are used to compute the length of a side in line 130. A similar technique is used for the circumscribed polygon. This step is done in line 140. The perimeters are computed in the same lines by multiplying the side by n: the number of sides in the polygon.

Line 155 terminates the program when the perimeters have converged within the accuracy of the machine.

Line 150 assumes a diameter of 2 for both polygons. Actually, the outer polygon has a diameter greater than 2.

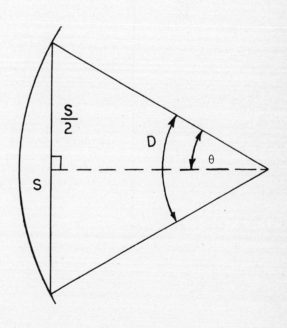

Write a computer program to accept the coordinates of the vertices of a triangle as input.

Have the program compute the coordinates of the intersection points of the medians, the perpendicular bisectors of the sides, the angle bisectors and the altitudes.

You may also want to type out the equations of some of these lines.

When you run the program, avoid involving lines which are vertical. These will have no slope. They will cause an error message because of the division by zero in your slope formula. You may wish to bypass this technique in favor of one which does not have this bug.

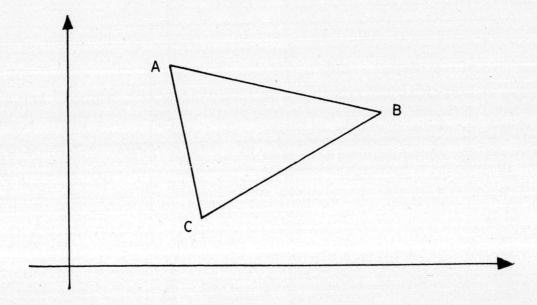

```
INPUT THE COORDINATES OF THE VERTICES OF A TRIANGLE
? 0,0,5,7,-2,4


MEDIANS MEET AT  2.8333333      3.4629629 THE CENTROID!
PERP BISECTORS OF THE SIDES MEET AT:
 2.2941176       3.6470589 THE CIRCUMCENTER!
ALTITUDES MEET AT -1.5882353     3.7058824 THE ORTHOCENTER!
```

```
10 PRINT 'INPUT THE COORDINATES OF THE VERTICES OF A TRIANGLE'
20 INPUT A,B,C,D,E,F
30 LET M1 = (A + C)/2
40 LET M2 = (B + D)/2
50 LET M3 = (E + C)/2
60 LET M4 = (F + D)/2
70 LET B1 = F-((M2-F)/(M1-E))*E
80 LET B2 = B-((M4-B)/(M1-E))*A
90 LET X = (B1-B2)/((M4-B-M2 + F)/(M1-E))
100 LET Y = ((M4-B)/(M1-E))*X + B2
110 PRINT
120 PRINT 'MEDIANS MEET AT ';X,Y;' THE CENTROID!'
125 IF D-B = 0 OR D-F = 0 THEN 400
130 LET S1 = (A-C)/(D-B)
140 LET S2 = (E-C)/(D-F)
150 LET K1 = M2-S1*M1
160 LET K2 = M4-S2*M3
170 LET X = (K1-K2)/(S2-S1)
180 LET Y = S1*X + K1
190 PRINT 'PERP BISECTORS OF THE SIDES MEET AT:'
200 PRINT X,Y;' THE CIRCUMCENTER!'
210 LET K1 = F-S1*E
220 LET K2 = B-S2*A
230 LET X = (K1-K2)/(S2-S1)
240 LET Y = S1*X + K1
250 PRINT 'ALTITUDES MEET AT ';X,Y;' THE ORTHOCENTER!'
390 GO TO 500
400 PRINT
410 PRINT 'ONE OF YOUR LINES HAS NO SLOPE!'
420 PRINT 'MY ALGORITHM INVOLVES SLOPE AVOID VERTICAL LINES'
430 GO TO10
500 END
```

ANALYSIS

Another approach to this problem would be feasible and interesting. The author chose to proceed by using slopes. This presents some problems if any vertical lines are encountered.

Lines 30 through 60 compute midpoints of the sides. Lines 70 and 80 solve a 2×2 system of equations for the slopes of the medians. Another set of equations is solved in lines 90 and 100. These result in the coordinates of the centroid.

Lines 130 and 140 compute the slope of the perpendicular bisector by simply taking the negative reciprocal of the given side. This slope is then used along with the midpoints previously computed to establish the equation of the perpendicular bisectors. This set of 2×2 is solved and printed out as the circumcenter. The same procedure is carried out for the altitudes, except that the coordinates of the vertex through which the altitude will pass is used and we know that it is parallel to the perpendicular bisector of the side so the slopes do not have to be recomputed.

Notice that only two medians, altitudes, etc. are needed to compute an intersection. A modification of this program could be used in a geometry course to provide empirical verification for that theorem.

31 Analysis of Triangular Coordinates

Prepare a program that will read in three sets of coordinates. Determine whether those coordinates are the vertices of a triangle.

If they are the sides of a triangle, type out what kind:

   Scalene, Isosceles, Equilateral.

Also check to see whether it is a right triangle. Have the computer draw a picture of the triangle using the TAB (X) function and finally compute and print the area of the triangle.

Be sure to use a general area formula and not one which only works for specific triangles.

References:

Kemeny and Kurtz, **Basic Programming,** pp. 47–50.

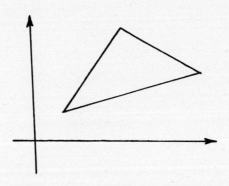

THIS PROGRAM WILL ACCEPT THE COORDINATES OF THE VERTICES
OF A TRIANGLE. IT WILL THEN COMPUTE THE AREA AND
THE TYPE OF TRIANGLE IN QUESTION!!!

INPUT THE COORDINATES OF THE VERTICES
? 0,0,3,0,0,3
THE AREA IS 4.5

IT'S A RIGHT TRIANGLE
THE TRIANGLE IS ISOSCELES!!
ANGLES IN DEGREES ARE 45 45 90

READY

RUNH
THIS PROGRAM WILL ACCEPT THE COORDINATES OF THE VERTICES
OF A TRIANGLE. IT WILL THEN COMPUTE THE AREA AND
THE TYPE OF TRIANGLE IN QUESTION!!!

INPUT THE COORDINATES OF THE VERTICES
? 0,1,4,5,-4,6
THE AREA IS 18

THE TRIANGLE IS SCALENE!
ANGLES IN DEGREES ARE 44.2152 52.125 83.6598

```
10 PRINT 'THIS PROGRAM WILL ACCEPT THE COORDINATES OF THE VERTICES'
20 PRINT 'OF A TRIANGLE. IT WILL THEN COMPUTE THE AREA AND'
30 PRINT 'THE TYPE OF TRIANGLE IN QUESTION!!!'
35 PRINT
40 PRINT 'INPUT THE COORDINATES OF THE VERTICES'
45 INPUT X1,Y1,X2,Y2,X3,Y3
50 LET S1 = SQR((Y2-Y1)↑2+(X2-X1)↑2)
55 LET S2 = SQR((Y3-Y1)↑2+(X3-X1)↑2)
57 LET S3 = SQR((Y3-Y2)↑2+(X3-X2)↑2)
60 LET S = (S1 + S2 + S3)/2
65 LET T = SQR(S*(S-S1)*(S-S2)*(S-S3))
70 PRINT 'THE AREA IS ';T
75 PRINT
77 IF S1↑2 + S2↑2 = S3↑2 THEN PRINT 'ITS A RIGHT TRIANGLE!'
78 IF S2↑2 + S3↑2 = S1↑2 THEN PRINT 'ITR A RIGHT TRIANGLE!'
79 IF S1↑2 + S3↑2 = S2↑2 THEN PRINT 'ITS A RIGHT TRIANGLE!'
80 LET P = 0
81 IF S1 < > S2 THEN 85
82 LET P = P + 1
85 IF S2 < > S3 THEN 105
90 LET P = P + 1
105 ON P + 1 GO TO 200,210,220
150 GO TO 300
200 PRINT 'THE TRIANGLE IS SCALENE!'
205 GO TO 300
210 PRINT 'THE TRIANGLE IS ISOSCELES!!'
215 GO TO 300
220 PRINT 'THE TRIANGLE IS EQUILATERAL!!!'
300 LET R = SQR((S-S1)*(S-S2)*(S-S3))/SQR(S)
310 LET A1 = (ATN(R/(S-S1))*2)*57.295779
320 LET A2 = (ATN(R/(S-S2))*2)*57.295779
330 LET A3 = (ATN(R/(S-S3))*2)*57.295779
340 PRINT 'ANGLES IN DEGREES ARE ';A1;A2;A3
400 END
```

ANALYSIS

The coordinates are used to find the lengths of the sides of the triangles in lines 50 through 57. The semi-perimeter for Hero's formula is computed in line 60. The area is computed in line 65.

Lines 77 through 79 test for a right triangle. These lines should be modified slightly. Truncation errors can cause right triangles to bypass this print-out as happened in the second run of the output. Notice the truncation in the area computation.

The test for the type of triangle is done by assigning points based on the number of sides which are equal. These points are stored in the variable P.

The angles are computed in lines 300 through 330. The formula is discussed in the ANALYSIS for the problem in this edition titled *Area of A Triangle by Coordinate Geometry Method.*

Write a program to compute the length of an arc for a given function.

Input the function in question and the limits (points) between which the length is to be computed. The input should be two numbers which will be *x*-values. Compute the *y*-values for those points. Have the computer divide the curve into a number of sections. Each section will be curved. Approximate that curve with a secant between the endpoints of the segment. The length of the secant can be found using coordinate geometry. It will be close to the arc length. Increase the number of intervals and sum them up until the answer is accurate to a specified degree.

Try to find a formula for arc length which can be used directly. Compare your approximation with the value as computed by formula.

References:

Donald Greenspan, **Introduction to Calculus,** pp. 206–209.

Ewind Moise, **Geometry,** pp. 371–406.

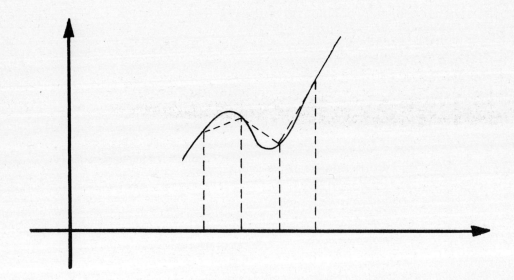

INTERVALS	APPROX ARC LEN
10	50.611573
20	50.674234
30	50.692439
40	50.698617
50	50.701355
60	50.702845
70	50.703736
80	50.704292
90	50.704713
100	50.704980

```
5 PRINT 'INTERVALS','APPROX ARC LEN'
10 DEF FNA(X)=X↑3-7*X+3
20 READ L,R
30 LET N=10
40 LET I=(R-L)/N
50 LET T=0
60 LET X1=L
70 LET X2=X1+I
80 LET Y1=FNA(X1)
90 LET Y2=FNA(X2)
100 LET D=SQR((X1-X2)↑2+(Y1-Y2)↑2)
110 LET T=T+D
120 LET X1=X2
130 LET X2=X2+I
140 IF X2<=R THEN 80
150 PRINT N,T
160 LET N=N+10
170 IF N<=100 THEN 40
180 DATA 0,4
200 END
```

ANALYSIS

Notice that even with only 10 intervals the computation of the length is fairly accurate. The procedure is to draw as many secants as there are intervals and then to compute and sum up these secant lengths.

Line 40 computes the size of each of the n intervals. Lines 80 and 90 establish the y values for the endpoint of each sub-interval and line 100 is the workhorse. It computes the length of each sub-interval (secant) by the coordinate geometry formula. The totals are incremented in line 110.

Line 150 prints the result including the partial sum. A further exercise would be to develop a formula for actual arc length and then to compare it with the approximations given by this program.

61

33 Sine and Cosine Tables by Computer

Ever wonder where sine and cosine tables come from? They are generated by tedious calculation using infinite series.

Use the series given below to generate a table for the sines and cosines of all angles between 0° and 90°.

Careful though, the variable x in both series must be in radians or the numbers will be meaningless. You'll have to have a subroutine to convert from degrees to radians and back. NO radians allowed in the output.

You choose the final form of the table. Try to get four places of accuracy. To avoid roundoff error, you may have to devise a way of retaining all the significant digits of the partial sums in an array. You may use the SIN and COS built-in functions to check accuracy only.

The series to be used:

$$\sin x = x - \frac{x^3}{3!} + \frac{x^5}{5!} - \frac{x^7}{7!} + \ldots$$

$$\cos x = 1 - \frac{x^2}{2!} + \frac{x^4}{4!} - \frac{x^6}{6!} + \ldots$$

References:

Other series representations may be used... They may be found in such books as:

CRC, **Standard Mathematical Tables,** p. 408.

Heinrich Dorrie, **100 Great Problems in Elementary Mathematics,** pp. 59–64.

DEGREES	SINE	COSINE	DEGREES
0	0	1	90
1	.017452	.999848	89
2	.0349	.999391	88
3	.052336	.99863	87
4	.069756	.997564	86
5	.087156	.996195	85
6	.104529	.994522	84
7	.121869	.992546	83
8	.139173	.990268	82
9	.156435	.987688	81
10	.173648	.984808	80
11	.190809	.981627	79
12	.207912	.978148	78
13	.224951	.97437	77
14	.241922	.970296	76
15	.258819	.965926	75
16	.275638	.961262	74
17	.292372	.956305	73
18	.309017	.951057	72
19	.325568	.945519	71
20	.34202	.939693	70
21	.358368	.93358	69
22	.374607	.927184	68
23	.390731	.920505	67
24	.406737	.913545	66
25	.422619	.906308	65
26	.438371	.898794	64
27	.453991	.891007	63
28	.469472	.882948	62
29	.48481	.87462	61
30	.5	.866025	60
31	.515038	.857167	59
32	.52992	.848048	58
33	.544639	.83867	57
34	.559193	.829037	56
35	.573577	.819152	55
36	.587786	.809017	54
37	.601816	.798635	53
38	.615662	.788011	52
39	.629321	.777146	51
40	.642788	.766044	50
41	.656059	.754709	49
42	.669131	.743145	48
43	.681999	.731353	47
44	.694659	.71934	46
45	.707107	.707106	45

DEGREES	SINE	COSINE	DEGREES

```
10 PRINT 'DEGREES', 'SINE', 'COSINE'
20 PRINT          _
22 FOR X = 0 TO 45
25 LET Y = X/57.295779
40 S = Y - (Y|3)/6 + (Y|5)/120 - (Y|7)/5040 + (Y|9)/362880 - (Y|11)/39916800
50 C = 1 - (Y|2)/2 + (Y|4)/24 - (Y|6)/720 + (Y|8)/40320 - (Y|10)/3628800
60 PRINT X,S,C,90-X
100 NEXT X
110 PRINT 'DEGREES', 'SINE', 'COSINE', 'DEGREES'
200 END
```

ANALYSIS

It is surprising to see that only five terms were needed for the series given to produce a value for the trig functions which was accurate to five places. The fact that the factorials had been input and not computed was a time-saver.

A bit more of a challenge could be provided by computing each term separately and then alternating its sign and keeping track of the partial sum.

Notice the use of the trig notion that the co-function of the complement is equal to the original angle's function.

Law of Sines—Ambiguous Case

Write a program which accepts as input two sides of a triangle and the angle opposite one of these sides.

The program should work for any type triangle *oblique* or *right*. Have the program type out the size of the third side along with how many triangles it is possible to construct with those dimensions. If the triangle is a right triangle, have the program say so.

Refer to the diagram below. Input A, B and π. The diagram should help you set the problem up. Use the right triangle which was constructed around the original. Involve a relationship with the X and Y as well as B and H.

Reference:

Julian Mancill, **Modern Analytical Trigonometry**, pp. 167–185.

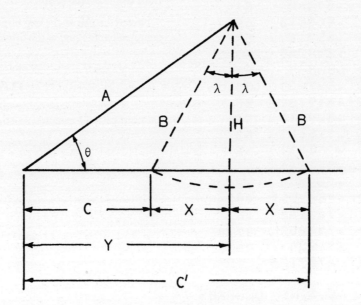

```
INPUT THE TWO SIDES AND ANGLE IN DEGREES
? 6,6,60
ONLY ONE TRIANGLE IS POSSIBLE
DIMENSIONS  6  6           6            60   DEGREES
ANY MORE TO BE DONE (YES OR NO)? YES


INPUT THE TWO SIDES AND ANGLE IN DEGREES
? 1,3,34
ONLY ONE TRIANGLE IS POSSIBLE
DIMENSIONS  1  3           3.77646      34   DEGREES
ANY MORE TO BE DONE (YES OR NO)? YES


INPUT THE TWO SIDES AND ANGLE IN DEGREES
? 1,23,67
ONLY ONE TRIANGLE IS POSSIBLE
DIMENSIONS  1  23          23.3723      67   DEGREES
ANY MORE TO BE DONE (YES OR NO)? YES.


INPUT THE TWO SIDES AND ANGLE IN DEGREES
? 9,10,135
ONLY ONE TRIANGLE IS POSSIBLE
DIMENSIONS  9  10          1.34966      135  DEGREES
ANY MORE TO BE DONE (YES OR NO)? NO
```

ANALYSIS

Given two sides of a triangle and an angle opposite one of them this program will tell how many triangles can be formed. It will also compute the remaining side.

Using the diagram in the student problem section, the following facts will prove useful.

Form a right triangle with the angle θ and side a. The sides h and y can be calculated from:

$$h = a \cdot \sin \theta$$
$$y = a \cdot \cos \theta$$

If $b < h$ no solution exists.

If $b = h$ only one solution exists, namely, the right triangle. The third side is y.

If $b > h$ the length of x is computed from

$$x = \sqrt{b^2 - h^2}$$

Within this framework two possibilities exist:

1) If $x < y$ then two solutions exist $(y + x)$ and $(y - x)$
2) If $x > y$ then only one solution exists: $(y + x)$

Line 40 converts from degrees to radians. Lines 50 and 60 compute h and y. The comparisons in lines 70 and 80 were specified because of truncation errors which are possible with trig functions.

The remainder of the program is fairly easy to follow if the diagram is used.

```
10 DIM A$(10)
20 PRINT
25 PRINT 'INPUT THE TWO SIDES AND THE ANGLE IN DEGREES'
30 INPUT A,B,T
40 LET R = T/57.295779
50 LET H = A*SIN(R)
60 LET Y = A*COS(R)
70 IF H-B > .01 THEN 90
80 IF B-H > .01 THEN 110
85 PRINT 'THE ONLY TRIANGLE POSSIBLE IS A RIGHT TRIANGLE'
87 PRINT 'DIMENSIONS ';A,B,Y,T;' DEGREES'
88 GO TO 150
90 PRINT 'NO TRIANGLE POSSIBLE!!'
100 GO TO 150
110 LET X = SQR(B↑2-H↑2)
120 IF X < Y THEN 140
126 PRINT 'ONLY ONE TRIANGLE IS POSSIBLE'
127 PRINT 'DIMENSIONS ';A,B,Y + X,T;' DEGREES'
130 GO TO 150
140 PRINT 'TWO TRIANGLES ARE POSSIBLE'
145 PRINT 'THEIR DIMENSIONS ARE -'
147 PRINT A,B,Y + X,T;' DEGREES'
148 PRINT A,B,Y-X,T;' DEGREES'
149 PRINT
150 PRINT 'ANY MORE TO BE DONE (YES OR NO)'
160 INPUT A$
170 IF A$ = 'YES' THEN 20
200 END
```

35   **Resolving a System of Vectors**

Physics students will remember the tedious calculations involved in working out the resultant of a given set of vectors.

Assume that all vectors, up to a total of ten, originate at the origin of a cartesian coordinate system. Using as input the magnitude and direction of each vector, write a program to compute and plot the magnitude and direction of the resultant.

You may wish to print out the intermediate components of each vector in tabular form, much as we did in the solution of such problems with paper and pencil.

An actual picture of the situation from the computer would be quite impressive also.

A typical system might look like this!

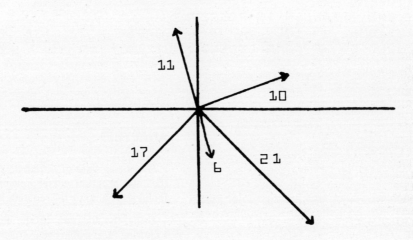

A typical system might look like this!

References:

Kemeny and Kurtz, **Basic Programming,** pp. 120–124.

```
THIS PROGRAM WILL RESOLVE A SYSTEM OF VECTORS!!

HOW MANY VECTORS TO BE RESOLVED
? 5
MAGNITUDE        ANGLE (D)        X-COMP           Y-COMP

   10              25           9.0630778        4.2261826
   11             100          -1.9101299       10.832885
   17             225         -12.020815       -12.020815
    6             280           1.0418889       -5.9088465
   21             315          14.849242       -14.849243

RESULTANT IS   20.868755 AT -58.114831 DEGREES
```

PLOT HAS BEEN OMITTED.

```
10 LET R = 57.295779
20 LET X1 = Y1 = 0
30 PRINT 'THIS PROGRAM WILL RESOLVE A SYSTEM OF VECTORS!!'
40 PRINT
45 PRINT 'HOW MANY VECTORS TO BE RESOLVED'
50 INPUT N
55 PRINT 'MAGNITUDE','ANGLE (D)','X-COMP','Y-COMP'
60 PRINT
70 FOR W = 1 TO N
80 READ M,A
85 LET X = M*COS(A/R)
90 LET Y = M*SIN(A/R)
95 PRINT M,A,X,Y
100 LET X1 = X1 + X
110 LET Y1 = Y1 + Y
120 NEXT W
130 LET R1 = SQR(X1↑2 + Y1↑2)
140 LET A1 = ATN(Y1/X1)*R
141 IF X1 < 0 THEN A1 = A1 + 180
145 PRINT
150 PRINT 'RESULTANT IS ';R1;' AT ';A1;' DEGREES'
160 PRINT
190 DATA 10,25,11,100,17,225,6,280,21,315
200 END
```

ANALYSIS

The program resolves the system by computing components and then summing them up. X components are computed in line 85. The R is to convert degrees to radians. The Y components are done in line 90.

Lines 100 and 110 sum the components. Then the right triangle for the components is solved for its hypotenuse in line 130 and the angle in line 140.

The negative angle in the output of course means that the angle should be turned clockwise from the x-axis.

One addition which is left to the reader is to print out the sum of both X and Y components before printing the solution.

67

36

Analysis of Projectile Motion

Projectile motion is one of the more interesting branches of physics. The tedious nature of the calculations, however, sometimes leaves students unconvinced of the foregoing assertion.

Given as input the muzzle velocity, angle of firing and other variables, program the computer to print out the range of the projectile along with the height to which it will rise, then have the computer print out a picture of the path (which we know will be a parabola).

Label both the vertical and horizontal axes with the proper units and dress the program up to be able to include two or more projectiles on one run.

Missile and missile intersection graphs are a possibility here.

Be sure the formulas used are correct for the units you introduce. You may wish to include air resistance in your study. You will then need to know more about the air such as temperature, density, etc.

References:

Isaac Asimov, **Understanding Physics: Motion, Sound and Heat.**

```
INPUT MUZZLE VELØCITY (FT/SEC) AND ANGLE (DEGREES)
? 210,30
MAXIMUM HEIGHT IS   172.15995 FEET
RANGE IS   1193.0366 FEET
TØTAL TIME AIRBØRNE IS   6.7599996 SECØNDS
SCALE ØF HEIGHT IS 1 SPACE =   3 FEET
          +++++++++++++++++++++++++++++++++++++++++++++++++++++++++++++++++++++
0         +
  36.373068+        *
  72.746136+              *
  109.1192 +                  *
  145.49227+                     *
  181.86534+                        *
  218.23841+                           *
  254.61147+                              *
  290.98454+                                *
  327.35761+                                  *
  363.73068+                                    *
  400.10375+                                     *
  436.47681+                                      *
  472.84988+                                       *
  509.22295+                                        *
  545.59602+                                        *
  581.96909+                                         *
  618.34216+                                         *
  654.71523+                                        *
  691.08829+                                        *
  727.46136+                                       *
  763.83442+                                      *
  800.20750+                                     *
  836.58057+                                   *
  872.95363+                                 *
  909.32670+                               *
  945.69976+                             *
  982.07283+                          *
  1018.4459+                       *
  1054.8190+                    *
  1091.192 +                 *
  1127.5651+             *
  1163.9382+         *
ANYMØRE (YES ØR NØ)? NØ
```

```
10 DIM X(126),K(2)
20 DEF FNA(X,Y,Z)=X*SIN(A*.01745329)*Z-16*Z↑2
30 PRINT 'INPUT MUZZLE VELOCITY (FT/SEC) AND ANGLE (DEGREES)'
40 F2=0
50 INPUT M,A
60 X(0)=0
70 FOR T=.2 TO 25 STEP .2
80 LET L=INT(5*T+.001)
90 X(L)=FNA(M,A,T)
100 IF X(L)<0 THEN 150
110 IF X(L)-X(L-1)<=0 AND F2=0 THEN 250 ELSE 350
120 K(1)=X(L-2)
130 W4=T-.4+J
140 GO TO 280
150 LET W=L-1
160 FOR J=.01 TO .2 STEP .01
170 K(1)=FNA(M,A,T-.2+J)
180 IF K(1)<0 THEN 200
190 NEXT J
200 PRINT 'MAXIMUM HEIGHT IS '; K; ' FEET'
210 LET Y=M*COS(A*.01745329)*(T-.2+J-.01)
220 PRINT 'RANGE IS ';Y;' FEET'
230 PRINT 'TOTAL TIME AIRBORNE IS ';T+J-.01;' SECONDS'
240 GO TO 360
250 IF X(L-1)-X(L-2)<X(L)-X(L-1)THEN 120
260 K(1)=X(L-1)
270 W4=T-.2+J
280 F2=1
290 FOR J=.01 TO .2 STEP .01
300 K(2)=FNA(M,A,W4)
310 IF K(1)-K(2)<=0 THEN 340
320 K(1)=K(2)
330 NEXT J
340 K=MAX(K(1),K(2))
350 NEXT T
360 IF K+10 >60 THEN 470 ELSE M1=1
370 FOR D=1 TO 62
380 PRINT TAB(10);'+';
390 NEXT D
400 PRINT '0';TAB(10);'+'
410 FOR N=1 TO W
420 PRINT.2*N*M*COS(A*.01745329);TAB(10);'+';TAB(X(N)*M1+10);'*'
430 NEXT N
440 PRINT 'ANYMORE (YES OR NO)';
450 INPUT A$
460 IF A$='YES' THEN 30 ELSE 500
470 M1=1/(INT(K/60+1))
480 PRINT 'SCALE OF HEIGHT IS 1 SPACE = ';1/M1;' FEET'
490 GO TO 370
500 END
```

ANALYSIS

This program computes the maximum height, range and time in flight of a projectile. Input includes the muzzle velocity and angle of firing. The flight of the projectile is graphed.

A vector analysis is used to calculate the height of the projectile at a given time (line 20), X is the muzzle velocity, A is the angle and Z is the time. The constant .01745329 changes degrees to radians.

The height of the projectile is calculated for each interval of 0.2 seconds in lines 70 throught 90. It is then stored in the subscripted variable X for later use in the graph.

When a change of direction is noted in the path, the time of the maximum height is calculated to the nearest 0.01 seconds (lines 250 through 330). The time of impact is calculated to the nearest hundredth also. It notes in lines 120 through 180 when the projectile's height is closest to zero.

Lines 360 through 430 plot the path. Lines 470 and 480 scale the drawing down so it will fit within the 72 character page width of the TTY.

70

Write a program to plot on the same set of axes any two pairs of the functions listed below.

Look over the situation to be sure that both your vertical and horizontal plot will include any points of intersection.

As a *special*, have the computer predict the points of intersection and label them with something other than the asterisks or other symbols used in the rest of the plot.

Go at least one full cycle for any function.

THE FUNCTIONS

1) sin and cosine
2) log and sin
3) $x^2 + 6x + 8$ and $y = 2x - 3$

Be sure the program is versatile enough to vary the horizontal and vertical spacing at the programmer's request.

Specify the units being used or better yet actually type them out as the axis for the plot.

You might do well to try a few of the library plot programs to see how they accomplish the task.

```
THIS PROGRAM WILL PLOT ANY TWO FUNCTIONS
TYPE THE FUNCTIONS IN BY USING LINES 100 AND 105
AND A DEF FNA AND DEF FNB STATEMENT
INPUT THE INTERVAL TO BE USED ON 'X' AND INCREMENT
? 0,6.3,0.3
INPUT INTERVAL ON 'Y' AND # OF SUBDIVISIONS
? -1,1,50
Y-AXIS: FROM -1 TO  1 STEP  .04
```

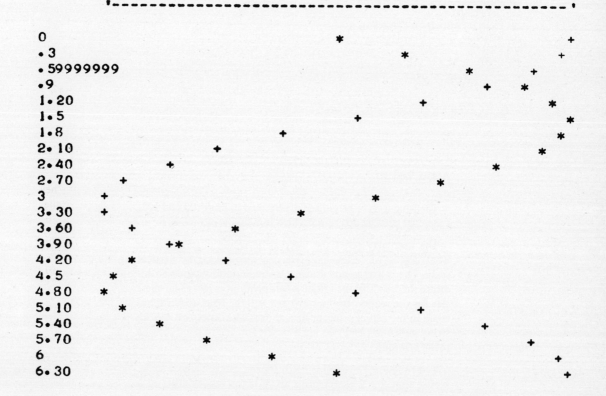

References:

Kemeny and Kurtz, **Basic Programming,** pp. 53–55.

```
10 PRINT 'THIS PROGRAM WILL PLOT ANY TWO FUNCTIONS'
20 PRINT 'TYPE THE FUNCTIONS IN BY USING LINES 100 AND 105'
30 PRINT 'AND A DEF FNA AND DEF FNB STATEMENT'
100 DEF FNA(X) = SIN(X)
105 DEF FNB(X) = COS(X)
110 DEF FNR(X) = INT(X + .5)
120 DEF FNX(X) = INT(100*X + .5)/100
125 PRINT 'INPUT THE INTERVAL TO BE USED ON "X" AND INCREMENT'
130 INPUT A,B,S
135 PRINT 'INPUT INTERVAL ON "Y" AND # OF SUBDIVISIONS'
140 INPUT C,D,N
150 IF N < = 50 THEN 180
160 PRINT '50 IS THE MOST YOU MAY USE'
170 GO TO 135
180 LET H = (D-C)/N
190 PRINT 'Y-AXIS: FROM ';C;' TO ';D;' STEP ';H
200 PRINT
210 PRINT TAB(8);'''';
220 FOR I = 1 TO N-1
230 PRINT '-';
240 NEXT I
250 PRINT ''''
255 PRINT
260 FOR X = A TO B STEP S
270 LET Y = FNA(X)
272 LET Y1 = 8 + FNR((Y-C)/H)
274 LET Y = FNB(X)
276 LET Y2 = 8 + FNR((Y-C)/H)
278 PRINT FNX(X);
280 IF Y2 < Y1 THEN 295
282 IF Y1 = Y2 THEN 290
285 PRINT TAB(Y1);'*';TAB(Y2);'+'
287 GO TO 300
290 PRINT TAB(Y1);'X'
292 GO TO 300
295 PRINT TAB(Y2);'+';TAB(Y1);'*'
300 NEXT X
305 PRINT
400 END
```

ANALYSIS

The functions to be plotted are defined in lines 100 and 105. Lines 280 and 282 determine which function is plotted first.

The user may select the interval he wants plotted and to what degree of refinement. He can also select the vertical scale. The x-axis is not drawn. A simple modification to indicate the x-axis in the initial y-axis printout is left as an exercise for the reader. In lines 210 and 250 the four apostrophes result in the actual printout of just one. The outer two are the string indicators and two must be used inside to produce one in the printout.

The function FNX (X) is used to print out the x-axis scale. Notice that line 290 will print an X if the curves cross at a point given on the graph. Line 282 sets this up.

38   A Quickie That May Take A While

Do this one by trial and error, but give the process some thought first.

Can you find a five digit number which when multiplied by four has its digits reversed?

Essentially what we want is a number *ABCDE* such that:

$$4 \times ABCDE = EDCBA$$

You'll need to develop a recognition algorithm that will know when the digits of a number are the reverse of the original.

An old thought comes to mind. When a number and its *mirror image* are subtracted, one from the other, the difference is always a multiple of nine. It's just a thought, not necessarily a hint.

21978 IS THE NUMBER!!!

```
10 FOR A = 1 TO 2
20 FOR B = 0 TO 9
30 FOR C = 0 TO 9
40 FOR D = 0 TO 9
50 FOR E = 0 TO 9
60 LET X = (10000*A) + (1000*B) + (100*C) + (10*D) + E
70 LET Y = (10000*E) + (1000*D) + (100*C) + (10*B) + A
80 IF 4*X < > Y THEN 100
85 PRINT
90 PRINT X;' IS THE NUMBER!!!'
100 NEXT E
110 NEXT D
120 NEXT C
130 NEXT B
140 NEXT A
200 END
```

ANALYSIS

Rather than take a number and separate it into its component digits, the approach here was the opposite. We generate the digits and recompose them into a number. That way the physical process of reversing the digits is just that: a physical process. The mathematics stays uncomplicated.

Statement 80 allows only the number with the property in question to be printed out.

73

Think of a Number

Many number games start with the phrase, *Think of a number!*

Many different algorithms have been used to do this. One of them is the Chinese Remainder Theorem. Think of a number less than 316. Write down the remainders when that number is divided by 5, 7 and 9. Using only those remainders the computer should be able to reconstruct the original number.

Research the Chinese Remainder Theorem and write a computer program to use it to find the number someone has thought of.

References:

William LeVeque, **Elementary Theory of Numbers**, p. 52.

```
THINK OF A POSITIVE INTEGER LESS THAN 316 !

DEVIDE YOUR NUMBER BY 5 WHAT IS THE REMAINDER? 3
DEVIDE YOUR NUMBER BY 7 WHAT IS THE REMAINDER? 4
DEVIDE YOUR NUMBER BY 9 WHAT IS THE REMAINDER? 7

YOUR NUMBER IS  88

THINK OF A POSITIVE INTEGER LESS THAN 316 !

DEVIDE YOUR NUMBER BY 5 WHAT IS THE REMAINDER? 0
DEVIDE YOUR NUMBER BY 7 WHAT IS THE REMAINDER? 0
DEVIDE YOUR NUMBER BY 9 WHAT IS THE REMAINDER? 3

YOUR NUMBER IS  210

THINK OF A POSITIVE INTEGER LESS THAN 316 !

DEVIDE YOUR NUMBER BY 5 WHAT IS THE REMAINDER? 1
DEVIDE YOUR NUMBER BY 7 WHAT IS THE REMAINDER? 1
DEVIDE YOUR NUMBER BY 9 WHAT IS THE REMAINDER? 1

YOUR NUMBER IS  1

THINK OF A POSITIVE INTEGER LESS THAN 316 !

DEVIDE YOUR NUMBER BY 5 WHAT IS THE REMAINDER? 1
DEVIDE YOUR NUMBER BY 7 WHAT IS THE REMAINDER? 4
DEVIDE YOUR NUMBER BY 9 WHAT IS THE REMAINDER? 2

YOUR NUMBER IS  11

THINK OF A POSITIVE INTEGER LESS THAN 316 !

DEVIDE YOUR NUMBER BY 5 WHAT IS THE REMAINDER? ↑P
STOP AT LINE 150
```

ANALYSIS

The algorithm is contained in line 180. Lines 130 through 160 store the values inputed by the user. R(5) is the remainder after the given number is divided by 5; R(7) is the remainder modulo 7; and R(9) is the remainder modulo 9.

These remainders are then multiplied by 126, 225 and 280 respectively and then summed up. That sum is divided by 315 and the remainder is the original number.

The reader is invited to research the algorithm using the references cited on the student problem page.

It is significant that $126 \equiv 1 \bmod 5$
$$225 \equiv 1 \bmod 7$$
$$280 \equiv 1 \bmod 9$$

```
100 PRINT
110 PRINT 'THINK OF A POSITIVE INTEGER LESS THAN 316 !'
120 PRINT
130 FOR I = 5 TO 9 STEP 2
140 PRINT 'DIVIDE YOUR NUMBER BY ';I;'
     WHAT IS THE REMAINDER':
150 INPUT R(I)
160 NEXT I
170 PRINT
180 PRINT 'YOUR NUMBER IS ';
     MOD((126*R(5) + 225*R(7) + 280*R(9)),315)
190 GO TO 100
200 END
```

Numbers: Perfect, Abundant and Deficient!

Write a program to test a given number to see whether it is PERFECT, ABUNDANT or DEFICIENT.

A number is:

Perfect: when the sum of the divisors of that number excluding the number itself equals the number in question.

Abundant: when the sum of the divisors exceeds the number.

Deficient: when the sum of the divisors is less than the number.

Examples:

$6 = 1 + 2 + 3$ and is PERFECT.

$12 \neq 1 + 2 + 3 + 4 + 6$. In fact, the sum exceeds 12. So 12 is abundant.

$10 \neq 1 + 2 + 5$. Here the sum falls short of 10. So 10 is deficient.

You must develop an algorithm to factor a number into its divisors. A number of them exist. Once you've completed that the rest is easy. Be sure to include 1 but exclude the number itself.

References:

Robert Wisner, **A Panorama of Numbers,** pp. 84–100.

A.H. Beiler, **Recreations in the Theory of Numbers,** pp. 11–26.

```
THIS PROGRAM WILL TAKE A NUMBER
AND COMPUTE THE SUM OF IT'S DIVISORS

INPUT THE NUMBER? 6
THE DIVISORS OF  6  ARE  1  2  3
 6  IS PERFECT!!

ANY MORE TO DO (YES OR NO)? YES

INPUT THE NUMBER? 12
THE DIVISORS OF  12  ARE  1  2  3  4  6
 12  IS ABUNDANT

ANY MORE TO DO (YES OR NO)? YES

INPUT THE NUMBER? 10
THE DIVISORS OF  10  ARE  1  2  5
 10  IS DEFICIENT.

ANY MORE TO DO (YES OR NO)? YES

INPUT THE NUMBER? 28
THE DIVISORS OF  28  ARE  1  2  4  7  14
 28  IS PERFECT!!

ANY MORE TO DO (YES OR NO)? NO
```

```
05 DIM A$(10)
10 PRINT 'THIS PROGRAM WILL TAKE A NUMBER '
20 PRINT 'AND COMPUTE THE SUM OF ITS DIVISORS'
30 PRINT
40 PRINT 'INPUT THE NUMBER ';
50 INPUT N
54 LET S=0
57 PRINT 'THE DIVISORS OF ';N;' ARE ';
60 FOR X=1 TO N-1
70 IF N/X < > INT(N/X) THEN 100
80 LET S=S+X
90 PRINT X;
100 NEXT X
110 PRINT
114 PRINT
115 PRINT S
120 IF S > N THEN 160
130 IF S < N THEN 190
140 PRINT N;' IS PERFECT!!'
150 GO TO 200
160 PRINT N;' IS ABUNDANT!!!!!!'
170 GO TO 200
190 PRINT N;' IS DEFICIENT'
200 PRINT
210 PRINT 'ANY MORE TO DO (Y OR N)'
220 INPUT A$
230 IF A$='Y' THEN 30
300 END
```

ANALYSIS

The entire problem in this program is to generate and sum up the divisors of a number.

This is done in the X loop in line 60 and ff. When X is a divisor it is added to the partial sum S and it is printed. The remainder of the program is conditional branches on the basis of the sum compared to the original number. Lines 120 and 130 effect this comparison.

This program could be used as the front end for a program to generate perfect numbers. Simply test and print out only perfect numbers.

41

Arithmetic Tables in Modulo N

Write a computer program to type out an addition table and a multiplication table modulo N, where N is any number up to 20.

Remember *modulo* means *remainder when divided by*.

For example:$17 \equiv 1 \bmod 4$

means that when 17 is divided by 4 the remainder is 1.

Have the computer type out two separate tables, one for multiplication and one for addition. You may use the MOD function if it is available. It might be more interesting to develop your own algorithm for computing the remainder. It can be done in two lines or less.

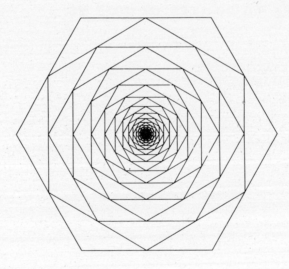

```
WHAT MODULUS? 7
* MOD( 7)
```

#	1	2	3	4	5	6
1↑	1	2	3	4	5	6
2↑	2	4	6	1	3	5
3↑	3	6	2	5	1	4
4↑	4	1	5	2	6	3
5↑	5	3	1	6	4	2
6↑	6	5	4	3	2	1

```
+ MOD( 7)
```

#	1	2	3	4	5	6
0	1	2	3	4	5	6
1	2	3	4	5	6	0
2	3	4	5	6	0	1
3	4	5	6	0	1	2
4	5	6	0	1	2	3
5	6	0	1	2	3	4
6	0	1	2	3	4	5

```
100 PRINT 'WHAT MODULUS';
110 INPUT N
120 PRINT '* MOD(';N;')'
130 LET L=-1
140 FOR X=1 TO N-1
150 IF L<0 THEN PRINT ' #';' ';
160 LET L=5
170 IF X>=0 THEN PRINT X;' ';
180 NEXT X
190 PRINT
200 PRINT '= = = = = = = = = = = =';
210 IF N=2 THEN 250
220 FOR W=1 TO N-3
230 PRINT '= = = = =';
240 NEXT W
250 PRINT
260 FOR I=1 TO N-1
270 FOR J=1 TO N-1
280 IF J=1 THEN PRINT I;'↑ ';
290 PRINT MOD(I*J,N);' ';
300 NEXT J
310 PRINT
320 NEXT I
330 PRINT
340 PRINT
350 PRINT
360 LET L=-1
370 PRINT '+ MOD(';N;')'
380 FOR Y=1 TO N-1
390 IF L<0 THEN PRINT ' #';' ';
400 LET L=5
410 IF Y>=1 THEN PRINT Y;' ';
420 NEXT Y
430 PRINT
440 PRINT '= = = = = = = = = = = =';
450 IF N=2 THEN 500
460 FOR W=1 TO N-3
470 PRINT '= = = = =';
480 NEXT W
490 PRINT
500 PRINT
510 FOR I=0 TO N-1
520 FOR J=0 TO N-1
530 PRINT MOD(I+J,N);' ';
540 NEXT J
550 PRINT
560 NEXT I
570 END
```

ANALYSIS

The program is lengthy because of the care taken to print the table in a neat format regardless of the size of the table.

The loops in 220 and 460 simply insure that the line drawn beneath the heading is the proper length. The meat of the program is the MOD function in lines 290 and 530. This function returns A modulo B when written as MOD (A,B). If this function is not available then the reader will have to use an alternative. One such is to take the greatest integer in A/B and subtract it from A after multiplying it by B. The result is the remainder after A is divided by B.

42

Gauss-Seidel Iterative Procedure

There is a technique for solving systems of equations which has probably occurred to some students.

For example when solving 3 equations in 3 unknowns, why not guess at the y and z, plug these in and generate x. Then take the new x which is now more than a guess and use the current z to generate a new y. Then take the generated x and y to generate a z. This process will iterate to a solution under certain conditions.

The iteration will not converge if the coefficient matrix of the original system is not diagonally dominant. This means if the system were

$$
\begin{array}{rcrcrcr}
4x & + & 2y & - & z & = & 11 \\
x & + & 7y & + & 2z & = & 16 \\
2x & - & 3y & - & 9z & = & 85
\end{array}
$$

this system would have a solution because:

$$
\begin{array}{rcll}
|4| \geq |2| + |-1| & \text{and} \\
|7| \geq |1| + |2| & \text{and} \\
|-9| \geq |-3| + |2| &
\end{array}
$$

In addition to these three conditions one of the above inequations must be a strict inequality. That is the left must be strictly greater than the right. Under these conditions the process will iterate to a solution.

Write a computer program to accept the equations which have been solved for the respective variables. You could even have the computer solve them. Input the coefficients and the initial guesses. Have the computer type out the approximations along the way. Include a test for convergence. Test to see whether a given value has changed in the last iteration.

```
GUESS Y & Z

?0,0
 8                       4                      1.10
 7.0333333               4.0148148              1.0048148
 6.9966667               4.0001645               .99968311
 6.9998394               3.9999469               .99997862
 7.0000106               3.9999988              1.0000009
```

```
10 DIM X(100),Y(100),Z(100)
20 READ A,B,C,D,E,F,G,H,U,J,K,L
25 LET I=U
30 IF ABS(A)<ABS(B)+ABS(C) THEN 200
40 IF ABS(F)<ABS(E)+ABS(G) THEN 200
50 IF ABS(K)<ABS(I)+ABS(J) THEN 200
60 IF ABS(A)>ABS(B)+ABS(C) OR ABS(F)>ABS(E)+ABS(G) THEN 210
70 IF ABS(K)>ABS(I)+ABS(J) THEN 210
200 PRINT 'CONVERGENCE UNLIKELY!!!'
205 GO TO 400
210 PRINT 'GUESS Y & Z'
220 INPUT Y(0),Z(0)
230 LET X(0)=0
240 FOR I=1 TO 100
250 LET X(I)=(D-B*Y(I-1)-C*Z(I-1))/A
260 LET Y(I)=(H-E*X(I)-G*Z(I-1))/F
270 LET Z(I)=(L-U*X(I)-J*Y(I))/K
280 PRINT X(I),Y(I),Z(I)
290 IF ABS(Y(I)-Y(I-1))<.0001 THEN STOP
300 NEXT I
350 DATA 3,1,-1,24,1,-9,1,-28,1,1,-10,1
400 END
```

ANALYSIS

The limitations on this algorithm are mentioned in the student problem section. They are tested in lines 30 through 70.

In line 220 the initial guesses for Y and Z are entered. The loop in 240 assumes that it will never take as many as 100 iterations. Lines 250 through 270 actually compute the values. These will remain the same regardless of the coefficients. The recycling of the updated variables is done by the subscripts. The current values are printed and the printout is terminated when the Y's differ by less than .0001 for two consecutive iterations. This was an arbitrary choice.

The program is limited to a 3×3 system.

Reference:

John Lee, **Numerical Analysis for Computers**, p. 165.

43

Self-Generating Integers

Self-generating integers are whole numbers whose separate digits, when factorialized (!) each by itself and added together, give the original number.

An example is not given here because there are only four such numbers known. There is one with three digits. However, an example of a trial could be: 1! + 2! + 3! is equal to 1 + 2 + 6 or 8 which is not equal to 123. Therefore, 123 is not a self-generating integer.

Write a computer program to find as many of these SGI's as you can. You must have a routine to factorialize an integer.

This is a time-consuming program so be as efficient as possible. Unless you have no time limit on CPU time, you may only be able to find a few or even one such SGI.

```
100 FOR A = 0 TO 9
110 FOR B = 0 TO 9
120 FOR C = 0 TO 9
125 FOR D = 0 TO 9
130 IF A < = 1 THEN X = 1
140 IF A = 2 THEN X = 2
150 IF A = 3 THEN X = 6
160 IF A = 4 THEN X = 24
170 IF A = 5 THEN X = 120
180 IF A = 6 THEN X = 720
190 IF A = 7 THEN X = 5040
200 IF A = 8 THEN X = 40320
210 IF A = 9 THEN X = 362880
220 IF B < = 1 THEN Y = 1
230 IF B = 2 THEN Y = 2
240 IF B = 3 THEN Y = 6
250 IF B = 4 THEN Y = 24
260 IF B = 5 THEN Y = 120
270 IF B = 6 THEN Y = 720
280 IF B = 7 THEN Y = 5040
290 IF B = 8 THEN Y = 40320
300 IF B = 9 THEN Y = 362880
310 IF C = <1 THEN Z = 1
320 IF C = 2 THEN Z = 2
330 IF C = 3 THEN Z = 6
340 IF C = 4 THEN Z = 24
350 IF C = 5 THEN Z = 120
360 IF C = 6 THEN Z = 720
370 IF C = 7 THEN Z = 5040
380 IF C = 8 THEN Z = 40320
390 IF C = 9 THEN Z = 362880
400 IF D < = 1 THEN W = 1
410 IF D = 2 THEN W = 2
420 IF D = 3 THEN W = 6
430 IF D = 4 THEN W = 24
440 IF D = 5 THEN W = 120
450 IF D = 6 THEN W = 720
460 IF D = 7 THEN W = 5040
470 IF D = 8 THEN W = 40320
480 IF D = 9 THEN W = 362880
500 IF 1000*A + 100*B + 10*C + D < > X + Y + Z + W THEN 550
510 PRINT A,B,C,D;'     WORK!!!!'
550 NEXT D
555 NEXT C
560 NEXT B
570 NEXT A
600 END
```

ANALYSIS

The program is designed to find only three and four digit self-generating integers.

Upon checking it is obvious that:

$$1! + 4! + 5! = 145$$

There appear to be no self-generating integers with four digits. There are however at least four more integers. The reader is invited to pursue the use of a DEF FN loop to compute the factorials. The great deal of checking necessary here makes the algorithm inefficient.

The literature on this subject is not voluminous. *The Journal of Recreational Mathematics* or the *Mathematical Gazette* have carried articles on the subject.

44

A Healthy List of Prime Numbers

Write a program to print in heirarchical order all prime numbers from 2 to 1000.

There is a classical algorithm for this process and you should know it.

You may not perform any multiplication or division in your program yet you will be able to generate all the primes called for in the exercise.

Print the primes horizontally, so as to conserve paper.

Remember 2 is a prime number.

HINT: Do not try to generate the primes directly. Eliminate the undesirables from your list. Let them fall through your *sieve* as it were.

References:

Robert Wisner, **A Panorama of Numbers,** pp. 67–71.

Kemeny and Kurtz, **Basic Programming,** pp. 60–63.

J.H. Caldwell, **Topics in Recreational Mathematics,** pp. 32–40.

2 3 5 7 11 13 17 19 23 29 31 37 41 43 47 53 59 61 67 71 73 79 83 89 97

101 103 107 109 113 127 131 137 139 149 151 157 163 167 173 179 181 191

193 197 199 211 223 227 229 233 239 241 251 257 263 269 271 277 281 283

293 307 311 313 317 331 337 347 349 353 359 367 373 379 383 389 397 401

409 419 421 431 433 439 443 449 457 461 463 467 479 487 491 499 503 509

521 523 541 547 557 563 569 571 577 587 593 599 601 607 613 617 619 631

641 643 647 653 659 661 673 677 683 691 701 709 719 727 733 739 743 751

757 761 769 773 787 797 809 811 821 823 827 829 839 853 857 859 863 877

881 883 887 907 911 919 929 937 941 947 953 967 971 977 983 991 997

```
05 DIM A(1000),B(200)
10 FOR X = 2 TO 1000
20 LET A(X) = 0
25 NEXT X
30 LET C = 0
35 LET S = SQR(1000)
40 FOR B = 2 TO 1000
50 IF A(B) < 0 THEN 100
60 LET C = C + 1
65 LET B(C) = B
70 IF B > S THEN 100
75 FOR X = B TO 1000 STEP B
80 LET A(X) = -1
90 NEXT X
100 NEXT B
110 PRINT
120 FOR X = 1 TO C
130 PRINT B(X);
140 NEXT X
200 END
```

ANALYSIS

The Sieve of Eratosthenes is the basis for this program. Line 10 begins the test loop. All numbers from 2 to 1000 are tested. Only those which are prime are printed.

The primes are stored in the array B(X). The A(X) array is a workspace. When A(X) is found to be divisible by a previous prime it is made negative and hence excluded from the list by line 50. Statement 65 tucks the next number not crossed out into the B(X) array to be printed later. The loop from 75 to 90 actually strikes out the multiples.

Line 70 insures that the algorithm doesn't do any wasteful computation. Any multiple which would be divisible by something greater than the SQR(1000) would have shown up when divided by the smaller factor.

The final loop from 120 to 140 prints the primes. The semicolon keeps the print element on the same line of type to conserve paper.

The output took only a quarter of a second. The program can be expanded to generate a large number of primes very easily. Just change all the 1000's to a larger number and save space in the *B* array for more primes.

It is interesting to study the density and frequency of primes with a large list.

45

Pascal's Triangle (The Challenging Way)

Write a program to generate piece by piece the rows of Pascal's triangle.

The triangle must be isosceles and the ones along the sides must be included.

Generate at least the first seven rows. Be sure enough space is allowed for double numbers, so that the triangle will not be unduly distorted.

As you know there are many ways to generate the triangle. Combinatorial methods, trigonometric methods, the binomial theorem. Choose the one you think is best suited and proceed.

The arithmetic generation of the triangle is to be discouraged. It is long and does not require much skill. We won't allow the first two rows to be entered so that the others can be generated by successive addition. There are too many other GOOD ways of generating it.

References:

See the pages on Pascal's triangle in the Appendix.

PASCALS TRIANGLE GENERATED!!

HEAH COME DA TRI-ANGLE!!!

```
                1
              1   1
            1   2   1
          1   3   3   1
        1   4   6   4   1
      1   5  10  10   5   1
    1   6  15  20  15   6   1
  1   7  21  35  35  21   7   1
1   8  28  56  70  56  28   8   1
```

```
100 DIM A(15),C(15,15)
110 PRINT
120 PRINT 'PASCALS TRIANGLE GENERATED!!'
130 PRINT
140 LET A(0)=1
150 FOR X=1 TO 10
160 LET A(X)=A(X-1)*X
170 NEXT X
180 LET B$='*'
190 FOR N=0 TO 8
200 FOR R=0 TO 8
210 IF N<R THEN 240
220 LET C(N,R)=A(N)/(A(R)*A(N-R))
230 IF C(N,R)< >0 THEN 240
240 NEXT R
250 NEXT N
260 PRINT
270 PRINT 'HEAH COME DA TRI-ANGLE!!!'
280 PRINT
290 FOR I=0 TO 8
300 FOR J=0 TO I
310 PRINT TAB(9-I);C(I,J);
320 NEXT J
330 PRINT
340 NEXT I
350 END
```

ANALYSIS

The number of combinations of n things taken r at a time will produce the binomial expansion coefficients of the expansion of $(X + Y)$ to the nth power.

The digits of the triangle were produced in this way. The loop for X at 150 computes the factorials needed in the formula for combinations. Line 220 computes the number of combinations and stores it in the array C.

The nested loops in 290 print the triangle. Suppression of trailing zeroes is done by the incrementation of those loops. Only as many elements as the number of the row are printed. The TAB function in line 310 causes the triangle to be printed in the triangular rather than column form. Slight modifications to statements 310 and the change in the upper limits of the loop would allow more rows to be generated.

46

A number which is prime and of the form $2^p - 1$ where p itself is prime is known as a Mersenne prime.

These numbers are useful in the study of perfect numbers. (See problem of that title). Each Mersenne prime of the form 2^p-1 produces an even perfect number of the form below and every even perfect number is of this form.

There are no known odd perfect numbers! Write a program to find several p's that yield Mersenne primes, and find the corresponding even perfect numbers.

Form of Perfect Number.... $2^{p-1}(2^p - 1)$

References:

Kathy Ruckstahl and Charles Wilford, **More Chips from the Mathematical Log,** pp. 36–39

Ball, **Mathematical Recreations and Essays.**

Kraitchik, **Mathematical Recreations,** pp. 70–73

PRIME	MERSENNE PRI	PERFECT NO.
2	3	6
3	7	28
5	31	49 6
7	127	8 128
13	8 19 1	33550336

```
TIME LIMIT EXCEEDED - PROGRAM STOPPED.
100 PRINT 'PRIME', 'MERSENNE PRI ', 'PERFECT NO.'
110 PRINT
120 READ P
130 LET M=2↑P-1
140 FOR X=2 TO (2↑P-2)
150 IF M/X<>INT(M/X) THEN 170
160 GO TO 110
170 NEXT X
180 PRINT P,2↑P-1,2↑(P-1)*(2↑P-1)
190 PRINT
200 GO TO 120
210 DATA 2,3,5,7,11,13,17,19
220 END
```

```
100 PRINT 'PRIME','MERSENNE PRI','PERFECT NO.'
110 PRINT
120 READ P
130 LET M=2↑P-1
140 FOR X=2 TO (2↑P-2)
150 IF M/X < > INT(M/X) THEN 170
160 GO TO 110
170 NEXT X
180 PRINT P,2↑P-1,2↑(P-1)*(2↑P-1)
190 PRINT
200 GO TO 120
210 DATA 2,3,5,7,11,13,17,19
220 END
```

ANALYSIS

This program is a simple plug-in. The data supplies the prime numbers. The recursion relationship is guaranteed to produce primes and perfect numbers when primes are plugged in as exponents.

The test for primeness is important because the variable M may not always be prime. The reader is invited to eliminate it and rerun the program.

47

Continued Fraction Analysis

A continued fraction is an ordinary fraction which has been rewritten.

For example: 7/11 could be written as follows

$$1/(11/7) = 1/[1 + (4/7)] = 1/[1 + 1/(7/4)] = \ldots.$$

Write a computer program to accept any rational number and convert it to continued fraction form. The output need not be in the format given above. Simple *slashes* (/) may be used to indicate division.

It is possible to have the output as it is given above. It is most difficult, however, and hardly worth the extra time.

Reference:

A. Ya Khinchin, **Continued Fractions.**

```
INPUT NUMERATOR & DENOMINATOR? 111,193
1/ 1+1/ 1+1/ 2+1/ 1+1/ 4+1/ 1+1/ 4+
```

```
100 PRINT 'INPUT NUMERATOR & DENOMINATOR';
110 INPUT N,D
120 LET A = INT(D/N)
125 IF A = 0 THEN 190
130 LET B = D-(A*N)
140 PRINT '1/';A;'+';
160 LET D = N
170 LET N = B
175 IF N = 0 THEN 190
180 GO TO 120
190 END
```

ANALYSIS

The algorithm by which the continued fractions are generated is a simple one. The fraction entered is flipped over, the quotient is printed as a denominator under *1. A* is the quotient. The remainder is then put in fractional form—lines 160 and 170 and the process is continued until a zero is encountered. The zero test—line 175—is done before dividing to avoid an error message.

This problem will successfully accomodate even the largest fractions. The teacher or student is invited to reformat the output. Clean up the printing and the trailing + sign.

This program can also be used in conjunction with another problem in this volume. *The Infinite Network of Resistances* problem involves a continued fraction. The current program could be used as a subroutine within that problem.

48

Extended Precision Division

Write a program to carry out a division until it repeats. All rational numbers (fractions) can be expressed as decimals. All rational numbers repeat sooner or later.

Take a long, hard look at the long division algorithm you are so familiar with. Program the computer to do exactly what you do when you divide.

When you get it working print out all the fractions with denominator *17*. Take a look at the pattern. Try some other prime denominators. Try to establish a pattern for these. Use the greatest integer function (INT) to devise a test for divisibility.

LEAST EXPONENTS TABLE FOR PRIME DENOMINATORS 3 TO 97

PRIME	EXPONENT		PRIME	EXPONENT
3	1		**47	46
**7	6		53	13
11	2		**59	58
13	6		**61	60
**17	16		67	33
**19	18		71	35
**23	22		73	8
**29	28		79	13
31	15		83	41
37	3		89	44
41	5		**97	96
43	21			

INPUT DENOMINATOR
? 13

```
 1/  13 = . 0 7 6 9 2 3 0 7 6 9 2 3
 2/  13 = . 1 5 3 8 4 6 1 5 3 8 4 6
 3/  13 = . 2 3 0 7 6 9 2 3 0 7 6 9
 4/  13 = . 3 0 7 6 9 2 3 0 7 6 9 2
 5/  13 = . 3 8 4 6 1 5 3 8 4 6 1 5
 6/  13 = . 4 6 1 5 3 8 4 6 1 5 3 8
 7/  13 = . 5 3 8 4 6 1 5 3 8 4 6 1
 8/  13 = . 6 1 5 3 8 4 6 1 5 3 8 4
 9/  13 = . 6 9 2 3 0 7 6 9 2 3 0 7
10/  13 = . 7 6 9 2 3 0 7 6 9 2 3 0
11/  13 = . 8 4 6 1 5 3 8 4 6 1 5 3
12/  13 = . 9 2 3 0 7 6 9 2 3 0 7 6
```

INPUT DENOMINATOR
? 17

```
 1/  17 = . 0 5 8 8 2 3 5 2 9 4 1 1 7 6 4 7
 2/  17 = . 1 1 7 6 4 7 0 5 8 8 2 3 5 2 9 4
 3/  17 = . 1 7 6 4 7 0 5 8 8 2 3 5 2 9 4 1
 4/  17 = . 2 3 5 2 9 4 1 1 7 6 4 7 0 5 8 8
 5/  17 = . 2 9 4 1 1 7 6 4 7 0 5 8 8 2 3 5
 6/  17 = . 3 5 2 9 4 1 1 7 6 4 7 0 5 8 8 2
 7/  17 = . 4 1 1 7 6 4 7 0 5 8 8 2 3 5 2 9
 8/  17 = . 4 7 0 5 8 8 2 3 5 2 9 4 1 1 7 6
 9/  17 = . 5 2 9 4 1 1 7 6 4 7 0 5 8 8 2 3
10/  17 = . 5 8 8 2 3 5 2 9 4 1 1 7 6 4 7 0
11/  17 = . 6 4 7 0 5 8 8 2 3 5 2 9 4 1 1 7
12/  17 = . 7 0 5 8 8 2 3 5 2 9 4 1 1 7 6 4
13/  17 = . 7 6 4 7 0 5 8 8 2 3 5 2 9 4 1 1
14/  17 = . 8 2 3 5 2 9 4 1 1 7 6 4 7 0 5 8
15/  17 = . 8 8 2 3 5 2 9 4 1 1 7 6 4 7 0 5
16/  17 = . 9 4 1 1 7 6 4 7 0 5 8 8 2 3 5 2
```

```
100 PRINT 'INPUT DENOMINATOR'
110 INPUT N
120 FOR T = 1 TO N-1
130 LET X = 0
140 PRINT
150 LET A = T
160 PRINT A;'/';N;' = .';
170 IF (A*10)/N < 1 THEN 240
180 LET A = A*10
190 PRINT INT(A/N);
200 LET X = X + 1
210 IF X > = N-1 THEN 290
220 LET A = A-INT(A/N)*N
230 GO TO 170
240 LET A = A*10
250 PRINT '0';
260 LET X = X + 1
270 IF X > = N-1 THEN 290
280 GO TO 170
290 NEXT T
300 END
```

```
PRIME DEN      LEAST EXP
##### ###      ##### ###

   3              1
   7 **           6
  11              2
  13              6
  17 **          16
  19 **          18
  23 **          22
  29 **          28
  31             15
  37              3
  41              5
  43             21
  47 **          46
  53             13
  59 **          58
  61 **          60
  67             33
  71             35
  73              8
  79             13
  87             28
  89             44
  97 **          96
```

```
90 PRINT
95 PRINT 'PRIME DEN','LEAST EXP'
97 PRINT '##### ###','##### ###'
99 PRINT
100 READ D
105 IF D = 5 THEN 100
110 X = 10
120 N = 1
130 X = MOD(X,D)
140 IF X = 1 THEN 180
150 X = X*10
160 N = N + 1
170 GO TO 130
180 IF N = D-1 THEN PRINT D;'**',N ELSE PRINT D,N
190 GO TO 100
260 DATA 3,5,7,11,13,17,19,23,29,31,37,41,43,47,53
270 DATA 59,61,67,71,73,79,87,89,97
272 DATA 179,181,191,193,197,199,211,223,227,229,233,239,241,251
280 END
```

ANALYSIS

This is an extended precision division routine. It works exactly as long division does. It should be run only with proper fractions. It will generate $n - 1$ places of all the fractions with denominator n.

Line 170 does most of the work. Line 190 does the printout. Zeroes are carried down from the dividend by line 240. Line 220 does the subtraction of the product of the partial quotient and the divisor.

The table that was given in the problem section is significant. If we look at 17 we see that we will have to raise 10 to the 16th power before a division by 17 will give a remainder of 1. This means that 16 is a primitive root of 17. This is why there are 16 places of repetition in the fractional expansion of denominators of 17. Notice also that if the 16 places are split the first eight are the nines-complement of the last eight. Notice further that each of the sixteen decimal expansions is simply a cyclic permutation of the other expansions.

For 13 the primitive root is 6. It will only repeat every 6 places. Those six will still be nines-complementary when split in half. Furthermore, the cyclic pattern here occurs in two groups because the quotient when 13 is divided by 6 is 2. The reader is encouraged to expand some of the other fractions to verify this intriguing relationship.

The program PRIMITIVE has been included to show the reader how it is possible to generate the table given in the student problem section. The reader is invited to follow the program through for several denominators. The algorithm used is clever and efficient. It computes the remainders for successive powers of 10 without actually doing any extended precision multiplications or divisions.

49   G. H. Hardy's Dull Number!

There is a story circulating about the famous British mathematician G. H. Hardy and his meeting with the bright young Indian mathematician Ramanujan.

Hardy told how he had ridden in a taxi with a number which he considered very dull. Upon hearing the number Ramanujan promptly replied how really interesting the number was after all. He claimed it was the smallest integer which could be written as the sum of two cubes in two different ways!

Write a program to find the number on Hardy's taxi.

Trial and error solutions are certainly permissible, a bit of historical research wouldn't hurt either.

Essentially what you're looking for is an integer I such that:

$$I = x^3 + y^3$$
$$\text{and}$$
$$I = a^3 + b^3$$

where x, y, a, and b are four different numbers.

Assume that x, y, a, b belong to the natural numbers

```
G. H. HARDY'S DULL NUMBER IS    1729
IT CAN BE WRITTEN AS ........    1 ↑3 +   12 ↑3
                 AND........     9 ↑3 +   10↑3
```

```
10 FOR X = 1 TO 30
20 FOR Y = 1 TO 30
30 FOR A = 1 TO 30
40 FOR B = 1 TO 30
50 IF X = A OR Y = A THEN 100
60 IF X↑3 + Y↑3 < > A↑3 + B↑3 THEN 100
70 PRINT 'G. H. HARDY''S DULL NUMBER IS ';X↑3 + Y↑3
80 PRINT 'IT CAN BE WRITTEN AS ....... ';X;' 13 + ';Y;' 13'
90 PRINT '          AND....... ';A;' 13 + ';B;'13'
95 GO TO 200
100 NEXT B
110 NEXT A
120 NEXT Y
130 NEXT X
200 END
```

ANALYSIS

The program is really rather trivial. The sheer power of the computer is illustrated here. One would never dream of solving a problem in number theory by such crude methods were it not for the incredible speed of the machine.

Lines 50 and 55 insure that the commutation of two identical elements is not printed out as a solution. It tests for a unique pair. It is an example of some of the logic functions in BASIC which make it such a powerful problem-solving language.

It took over 2 seconds to test less than 30^4 possibilities.

Multiplication Tables in Several Bases

Prepare a program that will construct a multiplication table for any base from 2 to 10.

Have the computer accept a number from an INPUT statement and from that construct the table.

HINT: Use a double subscripted variable and keep in mind that element $A(m,n) = A(m,n)$.

References:

Robert Wisner, **A Panorama of Numbers,** pp. 10–18.

A. H. Beiler, **Recreations in the Theory of Numbers.,** pp. 67–72.

Aaron Bakst, **Mathematical Puzzles and Pastimes,** pp. 43–55.

$$4_5 \times 2_5 = 13_5$$

51 Amicable Numbers

Some numbers possess no status whatsoever, but when related to other numbers they become famous.

Such a number is 220. It doesn't appear to be unusual at all. In fact if we add up all of its integral divisors excluding the number itself we get:

$$1 + 2 + 4 + 5 + 10 + 11 + 20 + 22 + 44 + 55 + 110 = 284$$

Nothing startling there, not yet. Now let's try another unassuming number, 284. If we do the same for it we get:

$$1 + 2 + 4 + 71 + 142 = 220$$

A bit more intriguing, isn't it? There seems to be a partnership here between 220 and 284; such number pairs are called *amicable* numbers. There are around 400 such number pairs known today. In 1750, Euler discovered 59 such pairs.

Devise a program to search out at least five more *amicable* pairs. There are at least five less than 10,000,000.

If in your research you find other pairs, they are allowable only if the computer program produces them in the normal course of its run.

You'll need a method of factoring into all the integral divisors, much as we suggested for perfect numbers. These two problems make an ideal tandem program. Be sure to exclude the number itself from the list of divisors.

References:

Robert J. Wisner, **A Panorama of Numbers,** pp. 101–103.

A. H. Beiler, **Recreations in the Theory of Numbers,** pp. 26–30.

Elvin J. Lee, **History and Discovery of Amicable Numbers,** p. 77.

Steve Rogowski, **Computer Clippings,** p. 2.

```
AMICABLE NUMBER SEARCH
  6 AND   6 ARE AMICABLE NUMBERS!!!

 28 AND  28 ARE AMICABLE NUMBERS!!!

220 AND  284 ARE AMICABLE NUMBERS!!!

284 AND  220 ARE AMICABLE NUMBERS!!!
```

```
10 PRINT 'AMICABLE NUMBER SEARCH '
20 FOR X = 5 TO 500
30 LET S = W = 0
40 FOR K = 1 TO X-1
50 IF X/K < > INT(X/K) THEN 100
60 LET S = S + K
100 NEXT K
110 LET T = S
120 FOR L = 1 TO T-1
130 IF T/L < > INT(T/L) THEN 200
140 LET W = W + L
200 NEXT L
210 IF W < > X THEN 250
220 PRINT X; ' AND ';S; ' ARE AMICABLE NUMBERS!!!'
230 PRINT
250 NEXT X
300 END
```

ANALYSIS

The fact that 6 and 28 were printed out as amicable numbers should not surprise the reader. The program factors a number into all its divisors. Actually, *factor* is a poor word! It sums up the divisors and then takes that sum and *factors* it. It then adds up the second group of divisors. If the second sum brings us back to the number we started with then we have an amicable number or a perfect number.

Actually if we talk in terms of cycles, a perfect number has cycle *1*; an amicable number has cycle *2*. It would be interesting to modify this program to print out numbers with cycle *3* which one might call *friendly numbers* or some such thing.

Some modifications which could be used to increase efficiency follow. Notice that line 50 tests to see if K divides X. Line 130 tests the sum of the divisors of X that is T. We could conceivably cut the looping time in half by not only saving the divisor but also the quotient. These are left as an exercise for the reader.

A more efficient algorithm for computing the sum of the divisors is available. It involves the prime factors of the number. It will take less computer time and hence allow for the generation of more amicable pairs. There are a total of 5 pairs less than 10,000.

Twin Prime Generator—Companion

Once you have figured out how to generate prime numbers it should be a simple matter to modify the program and print out a set of *twin primes*. *Twin primes* are two numbers both of which are prime and which differ by two.

There is only one even prime number and that is 2. All other primes are odd. If two consecutive prime numbers are found they are called *twin primes*.

For example, 3 and 5 are twin primes, so are 17 and 19.

Have your program *split out* all such pairs less than 2000.

```
TWIN**PRIMES
```

3	5	5	7	11	13	17	19	29	31
41	43	59	61	71	73	101	103	107	109
137	139	149	151	179	181	191	193	197	199
227	229	239	241	269	271	281	283	311	313
347	349	419	421	431	433	461	463	521	523
569	571	599	601	617	619	641	643	659	661
809	811	821	823	827	829	857	859	881	883
1019	1021	1031	1033	1049	1051	1061	1063	1091	1093
1151	1153	1229	1231	1277	1279	1289	1291	1301	1303
1319	1321	1427	1429	1451	1453	1481	1483	1487	1489
1607	1609	1619	1621	1667	1669	1697	1699	1721	1723
1787	1789	1871	1873	1877	1879	1931	1933	1949	1951
1997	1999								

```
10 DIM A(2000),B(400)
20 FOR X = 2 TO 2000
25 LET A(X) = 0
30 NEXT X
35 LET C = 0
40 LET S = SQR(2000)
42 FOR B = 2 TO 2000
50 IF A(B) < 0 THEN 100
60 LET C = C + 1
65 LET B(C) = B
70 IF B > S THEN 100
74 FOR X = B TO 2000 STEP B
80 LET A(X) = -1
90 NEXT X
100 NEXT B
110 PRINT
115 PRINT '        TWIN**PRIMES'
117 PRINT
120 FOR X = 2 TO C
125 IF B(X)-B(X-1) < > 2 THEN 140
130 PRINT B(X-1);' ';B(X),
140 NEXT X
200 END
```

ANALYSIS

There are surprisingly few twin primes less than 2000. The algorithm is not quite as efficient as the one given for primes.

The program is essentially the same program which generated the list of primes. One statement, 80, was added to test for twins.

A more complete list could lead to a study of the density and frequency of primes. The topic is covered at length in an excellent little volume called *A Panorama of Numbers*. It is referenced in the bibliography at the end of this book.

The Inverse of a Matrix

Write a program to invert an $n \times n$ matrix where n is no bigger than five.

You may not use the MAT INV statements in the program for anything but a check.

We have developed several algorithms for inverting a matrix, any one of these are computer compatible. You might want to check the inverse by using it to multiply the original matrix and seeing if the identity matrix results. This is quite easy to do with the series of MAT statement available in BASIC.

References:

Kemeny and Kurtz, **Basic Programming.**

Calingaert, **Principles of Computation,** pp. 143–155.

Kenneth Loewen, **More Chips from the Mathematical Log,** p. 70.

```
100 PRINT 'INPUT RANK OF MATRIX';
50 REM BE SURE TO MODIFY INV() COMMAND AND DIMENSION
55 REM TO ASSURE PROPER RUN.....
110 INPUT N
115 PRINT
120 PRINT 'THE MATRIX IS A ';N;' BY ';N
125 LET W = (N-1)↑2
126 DIM D(100)
130 MAT B = DIM(N-1,N-1)
140 MAT A = DIM(N,N)
150 PRINT
160 MAT READ A(N,N)
170 FOR R = 1 TO N
180 FOR C = 1 TO N
190 PRINT A(R,C);'    ';
200 NEXT C
210 PRINT
220 NEXT R
230 MAT S = INV(A)
240 LET D2 = DET
250 FOR R = 1 TO N
260 FOR C = 1 TO N
270 L = 1
280 FOR I = 1 TO N
290 FOR J = 1 TO N
300 IF R = I THEN 340
302 IF J = C THEN 340
310 D(L) = A(I,J)
320 L = L + 1
340 NEXT J
350 NEXT I
360 L = 1
370 FOR X = 1 TO N-1
380 FOR Y = 1 TO N-1
390 B(X,Y) = D(L)
400 L = L + 1
410 NEXT Y
420 NEXT X
430 MAT T = INV(B)
440 LET D = DET
450 LET C(C,R) = (SGN((-1)↑(R + C))*D)/D2
460 NEXT C
470 NEXT R
475 PRINT
480 PRINT 'INVERSE IS :'
485 PRINT
490 FOR R = 1 TO N
500 FOR C = 1 TO N
510 PRINT C(R,C),
520 NEXT C
530 PRINT
540 NEXT R
550 DATA 2,3,-1,1,2,1,-1,-1,3
560 END
```

ANALYSIS

The computation of the inverse is done by a method described in the Schaum's Outline Series volume entitled *MATRICES*.

Essentially the matrix to be inverted is tested for invertibility. If its determinant is zero it cannot be inverted. If the determinant is non-zero it is saved in line 240. Notice that in line 230 the INVERSE of *A* was taken. This appears to be against the rules set down in the problem section. However, on this system (RTB—UNIVAC) the DET function for computing the determinant can only operate on the most recently inverted matrix. It has no argument and hence INV(A) caused line 240 to store the determinant of *A* in *D2*.

The computation of the inverse is essentially the same as the internal subroutine used by BASIC to compute inverses with the INV function. The adjoint of matrix *A* is computed in lines 250 through 470.

The adjoint is found by deleting the row and column of matrix *A* and substituting the value of the minor for these rows multiplied by (–1) raised to the *R* + *C* power where *R* = row and *C* = column. The matrix thus obtained is transposed. This is the adjoint. The deletion of the row and column in question is done by the test in 300 and 302.

The determinant array is stored in *B* in line 390. Its determinant is computed in 440. The transpose is taken in 450 by simply switching the subscripts.

The final step of dividing each element of the adjoint by the determinant of *A* is also done in line 450. The printout is done in the last series of nested loops beginning on line 490.

54 Magic Squares by Computer

A *Magic Square* is an array of numbers with just as many rows as columns. The sum of any row, column or diagonal is always the same. No number may be used more than once in constructing the array.

Write a computer program to generate magic squares up to 12 × 12. Let the user specify the size of the square. The sum to which all lines converge may be any positive integer. It could possibly be selected at random based upon a starting point which the user specifies.

The magic square shown below is remarkable in that it sums to 34 not only for all rows, columns and diagonals, but also for corner arrays, the center square of four and many more. It is attributed to Albrecht Durer and appeared in one of his paintings in the year 1514, as is indicated by the center squares on the very bottom line of the painting.

16	3	2	13
5	10	11	8
9	6	7	12
4	15	14	1

```
TYPE AN ODD NUMBER < 12 AND A RANDOM NUMBER < 800!
? 3, 76
N =  3            ====  SUM =   240
   79   84   77

   78   80   82

   83   76   81
```

```
TYPE AN ODD NUMBER < 12 AND A RANDOM NUMBER < 800!
? 9,789
N = 9          ==== SUM =  7461
```

825	866	817	858	809	850	801	842	793
794	826	867	818	859	810	851	802	834
835	795	827	868	819	860	811	843	803
804	836	796	828	869	820	852	812	844
845	805	837	797	829	861	821	853	813
814	846	806	838	789	830	862	822	854
855	815	847	798	839	790	831	863	823
824	856	807	848	799	840	791	832	864
865	816	857	808	849	800	841	792	833

```
TYPE AN ODD NUMBER < 12 AND A RANDOM NUMBER < 800!
? STOP
PROGRAM STOPPED.
```

```
100 DIM L(13,13)                  310 GO TO 190
110 GOSUB 340                     320 FNA = ((N*N + 2*P-1)*N)/2
120 GO TO 110                     330 FNEND
130 PRINT                         340 PRINT
140 DEF FNA(N,P,L)                350 PRINT 'TYPE AN ODD NUMBER < 12 AND A RANDOM NUMBER < 800!'
150 K = P                         360 INPUT N,P
160 J = INT((N + 1)/2)            370 N = MIN(11,MAX(3,N))
170 I = J + 1                     380 P = MAX(1,MIN(799,P))
180 M = 1                         390 S = FNA(N,P,L)
190 L(I,J) = K                    400 PRINT 'N = ';N,' = = = = SUM = ';S
200 K = K + 1                     410 PRINT
210 M = M + 1                     420 FOR I = 1 TO N
220 IF K = N*N + P THEN 320       430 PRINT
230 IF M < = N THEN 270           440 PRINT
240 I = I + 2                     450 FOR J = 1 TO N
250 IF I > N THEN I = I-N         460 PRINT TAB(5*J-2.001-LGT(L(I,J)));L(I,J);
260 GO TO 180                     470 NEXT J
270 I = I + 1                     490 NEXT I
280 J = J + 1                     500 PRINT
290 IF I > N THEN I = I-N         510 RETURN
300 IF J > N THEN J = J-N         520 END
```

ANALYSIS

The best way to follow this program is to trace a given route with a certain input.

The multiple line DEF from 140 to 330 actually generates the elements of the square. It uses the variable P entered by the user to initialize the array. The sum will depend on the choice of P. The important statement in the upper portion is line 320.

Only odd numbered arrays are produced. Line 460 assures that the array will be square and in a column oriented format.

Trace the program a couple of times to get the feel of the algorithm.

55

Write a program to compute the day of the week a given date occurred. Input should include the month, day and year in any order you specify. You need not input the month alphanumerically.

Many useful algorithms have been developed to serve as perpetual calendars. Some research is required on your part.

Set an upper and lower limit on the years. Remember Pope Gregory gave us our current calendar sometime during the fifteenth century. Things were really mixed up before then.

TODAY IS
JANUARY
2
MONDAY

References:

Kemeny and Kurtz, **BASIC Programming**, p. 36.

W. Ball, **Mathematical Recreations and Essays**, p. 449.

Aaron Bakst, **Mathematical Puzzles and Pastimes**, pp. 84–96.

Martin Gardner, **Mathematical Games**, *Scientific American*, May 1967.

I AM A PERPETUAL CALENDAR

I WILL CALCULATE THE DAY OF THE WEEK AN EVENT
TRANSPIRED, IF GIVEN THE DATE!!

USE THE CHART BELOW FOR MONTH NUMBERS

JAN-1 FEB-4 MAR-4 APR-0 MAY-2 JUN-5
JUL-0 AUG-3 SEP-6 OCT-1 NOV-4 DEC-6
FOR JAN AND FEB OF LEAP YEARS GO BACK ONE DAY
FOR DATES IN THE 1800'S GO AHEAD TWO DAYS
INPUT THE MONTH,DAY AND YEAR AND HIT RETURN
? 0, 12, 1973
SOCK IT TO THURSDAY

INPUT THE MONTH,DAY AND YEAR AND HIT RETURN
? 3, 17, 1968
HAPPENED ON A WEEKEND - SATURDAY TO BE EXACT!
INPUT THE MONTH,DAY AND YEAR AND HIT RETURN
? 3, 29, 1946
SOCK IT TO THURSDAY
INPUT THE MONTH,DAY AND YEAR AND HIT RETURN
? STOP
 PROGRAM STOPPED.

```
10 PRINT 'I AM A PERPETUAL CALENDAR'
20 PRINT
30 PRINT 'I WILL CALCULATE THE DAY OF THE WEEK AN EVENT'
40 PRINT 'TRANSPIRED, IF GIVEN THE DATE!!'
50 PRINT
60 PRINT 'USE THE CHART BELOW FOR MONTH NUMBERS'
70 PRINT
75 PRINT 'JAN-1 FEB-4 MAR-4 APR-0 MAY-2 JUN-5'
80 PRINT 'JUL-0 AUG-3 SEP-6 OCT-1 NOV-4 DEC-6'
81 PRINT 'FOR JAN AND FEB OF LEAP YEARS GO BACK ONE DAY'
82 PRINT 'FOR DATES IN THE 1800''S GO AHEAD TWO DAYS'
85 PRINT
90 PRINT
91 PRINT 'INPUT THE MONTH,DAY AND YEAR AND HIT RETURN'
95 INPUT A,B,C
100 IF C<1900 THEN *+2
110 GO TO *+2
120 LET C=C+100
130 LET Y=INT((C-1900)/12)
140 LET X=(C-1900)-(Y*12)
150 LET Z=INT(X/4)
160 LET W=INT((X+Y+Z)/7)
170 LET T=(X+Y+Z)-(7*W)
180 LET U=T+A
190 LET V=INT(U/7)
200 LET S=U-(7*V)
210 LET R=S+B
220 LET P=INT(R/7)
230 LET O=R-(7*P)
240 IF O=0 THEN *+2
250 ON O GO TO 300,302,304,306,308,310
260 PRINT 'HAPPENED ON A WEEKEND - SATURDAY TO BE EXACT!'
270 GO TO 90
300 PRINT 'THE DAY WAS A SUNDAY'
301 GO TO 90
302 PRINT 'MONDAY''S THE DAY!'
303 GO TO 90
304 PRINT 'WOULD YOU BELIEVE A TUESDAY'
305 GO TO 90
306 PRINT 'WEDNESDAY''S GUILTY'
307 GO TO 90
308 PRINT 'SOCK IT TO THURSDAY'
309 GO TO 90
310 PRINT 'THANK GOD IT WAS A FRIDAY!!'
320 GO TO 90
400 END
```

```
INPUT MONTH,DAY,YEAR IN NUMERIC FORM
? 1,18,1975

THE DAY OF THE WEEK WAS (OR WILL BE) SATURDAY


ANYTHING ELSE (YES OR NO)? YES
INPUT MONTH,DAY,YEAR IN NUMERIC FORM
? 8,17,1968

THE DAY OF THE WEEK WAS (OR WILL BE) SATURDAY


ANYTHING ELSE (YES OR NO)? YES
INPUT MONTH,DAY,YEAR IN NUMERIC FORM
? 8,29,1946

THE DAY OF THE WEEK WAS (OR WILL BE) THURSDAY


ANYTHING ELSE (YES OR NO)? YES
INPUT MONTH,DAY,YEAR IN NUMERIC FORM
? 2,3,1967

THE DAY OF THE WEEK WAS (OR WILL BE) FRIDAY


ANYTHING ELSE (YES OR NO)? NO




  5 DIM C$(6)=9
 10 FOR K=0 TO 6
 20 READ C$(K)
 30 NEXT K
 40 PRINT "INPUT MONTH,DAY,YEAR IN NUMERIC FORM"
 50 INPUT M,D,Y
 55 PRINT
 60 LET L=0
 70 LET X1=INT((Y-1)/4)
 80 LET X2=INT((Y-1)/100)
 90 LET X3=INT((Y-1)/400)
100 IF INT(Y/400)=Y/400 THEN L=1
110 IF INT(Y/4)#Y/4 THEN 120
115 IF INT(Y/100)#Y/100 THEN L=1
120 LET T=(M-1)*31
130 IF M>9 THEN T=T-6
140 IF M<=6 THEN 150
```

```
145 IF M<=9 THEN T=T-5
150 IF M<=4 THEN 160
155 IF M<=6 THEN T=T-4
160 IF M<=2 THEN 170
165 IF M<=4 THEN T=T-3
170 LET K=T+L+Y+D+X1-X2+X3
180 LET W=K-(INT(K/7)*7)
190 PRINT "THE DAY OF THE WEEK WAS (OR WILL BE) ";CS(W)
193 PRINT
197 PRINT
200 PRINT "ANYTHING ELSE (YES OR NO)";
210 INPUT A$
220 IF A$="YES" THEN 40
230 DATA SATURDAY,SUNDAY,MONDAY,TUESDAY,WEDNESDAY,THURSDAY,FRIDAY
240 END
```

ANALYSIS

There are many algorithms for calendar computation. The one presented here is by no means the most powerful. It has limitations with respect to leap years and centuries previous to the twentieth.

The program works as follows: consider the date July 19, 1973. The algorithm takes only the *73* (line 130), then takes the quotient when 73 is divided by 12. This would give 6 which is stored in Y. The remainder is 1 and is stored in X. Then the quotient of X divided by 4 is retained as Z, and the three X, Y and Z are summed. The remainder modulo 7 is retained from this sum and added to the month number from the table. For July it is 0. The remainder in this case would be 0 when X, Y and Z are divided by 7. This is added to the month number to give 0. The remainder modulo 7 is taken again giving 0. The day is added to that giving 19 in this example. This result is taken modulo 7 again giving 5 which corresponds to Thursday. That is the correct day. The days run from Saturday which is 0 through Friday which is 6.

A command MOD(X,Y) would shorten this program. It is not available in all versions of BASIC. It returns the remainder after X is divided by Y.

The month numbers selected (lines 75 and 80) were based upon an algorithm described in Martin Gardner's column as given in the references with the problem description. It appeared in May 1967 on page 136.

A second program using a different algorithm is included for the reader's convenience.

The challenge to find an arithmetic progression containing exactly one-hundred terms, all of them distinct primes has so far eluded mathematicians.

The longest at this writing consists of only twelve terms with the initial term 23143 and a difference of 30030. It was discovered by W. A. Golubiev.

It would not be necessary here to break the record of twelve. Just find another sequence with at least seven terms, all of which are distinct primes. The March 1964 issue of *Scientific American* has a column about patterns in primes which may be helpful.

References:

W. Sierpinski, **Some Unsolved Problems of Arithmetic**, *Twenty-eighth Yearbook of the NCTM*, p. 211.

Robert Wisner, **A Panorama of Numbers**, pp. 118–138.

Albert Beiler, **Recreations in the Theory of Numbers**, pp. 39–48.

57

Can you find numbers of the form:

$$n \cdot 2^n + 1$$

called Cullen numbers where a value for $n > 1$ produces a prime. Very recently it was shown that the least such prime was for $n = 141$.

Not only would you have to develop an algorithm to compute the exact digits of the number but you would also have to test its primeness. This is not a trivial matter when numbers become as large as 2^{141}.

References:

W. Sierpinski, **Some Unsolved Problems of Arithmetic**, *Twenty-eighth Yearbook of the NCTM*, p. 211.

Ibid., pp. 34–45.

Robert Wisner, **A Panorama of Numbers**, pp. 118–138.

N factorial, usually written *N!* may be defined as the product of the first *N* integers.

The beginning of a table for *N* factorial would look like this:

N	N factorial
1	1
2	2
3	6
4	24
5	120
6	720
7	5040
8	40320
9	362880
10	3628800

The problem is to carry this table out to 40 factorial, while still retaining all digits of accuracy. *N!* increases quite rapidly as *N* grows larger. A single variable in your program will not be able to contain all the digits.

You will have to devise a scheme to store the digits of *N!* in an array, one or two digits per element of the array. Then you will have to come up with a way of multiplying this array which represents a single large integer by *N* + 1 to obtain the next factorial.

Continue your table up to 40!, or until one line of output is filled, whichever comes first. The following relationship may prove helpful, it is called *Stirling's Approximation*:

$$\sqrt{2n\pi} \cdot (n/e)^n < n! < \sqrt{2N\pi}(n/e)^n \cdot 1 + \left(\frac{1}{12n-1}\right)$$

where $\pi = 3.14159265....$ and $e = 2.71828....$

Reference:

Steve Rogowski, **Computer Clippings**, pp. 8 and 171.

? 8

  0 4 0 3 2 0

? 16

  0 2 0 9 2 2 7 8 9 8 8 8 0 0 0

? 35

  0 1 0 3 3 3 1 4 7 9 6 6 3 8 6 1 4 4 9 2 9 6 6 6 6 5 1 3 3 7 5 2 3 2 0 0
  0 0 0 0 0 0

EXACTO

```
100 DIM S(1000)
105 PRINT 'TYPE IN # TO BE FACTORIALIZED';
108 INPUT N
110 LET S(1) = 1
120 LET M = 1
130 LET C = 1000
140 FOR X = 1 TO N
150 GOSUB 200
155 NEXT X
160 PRINT
170 PRINT ' '
180 FOR Z = 1 TO M
185 GOSUB 500
190 NEXT Z
195 STOP
200 FOR I = 1 TO M
210 GOSUB 300
220 NEXT I
230 FOR I = 1 TO M
240 GOSUB 400
250 NEXT I
260 RETURN
300 LET S(I) = S(I)*X
310 IF I+1-M < = 0 THEN 370
330 LET I = M + 1
370 RETURN
400 IF S(I)-C < 0 THEN 470
410 LET Z = INT(S(I)/C)
420 LET S(I) = S(I)-C*Z
425 LET S(I + 1) = S(I + 1) + Z
430 IF I-M < 0 THEN 470
440 LET M = I + 1
470 RETURN
500 LET I = M + 1-Z
505 LET L = 2
510 IF (7*INT(Z/7-Z)) < 0 THEN 520
515 LET L = 1
520 FOR J = 1 TO 3
530 GOSUB 600
540 NEXT J
550 RETURN
600 LET X = INT(S(I)/10↑(3-J))
620 PRINT X;
630 LET S(I) = S(I)-10↑(3-J)*X
640 RETURN
700 END
```

1	1
2	2
3	6
4	24
5	120
6	720
7	5040
8	40320
9	362880
10	3628800
11	39916800
12	479001600
13	6227020800
14	87178291200
15	1307674368000
16	20922789888000
17	355687428096000
18	6402373705728000
19	121645100408832000
20	2432902008176640000
21	51090942171709440000
22	1124000727777607680000
23	25852016738884976640000
24	620448401733239439360000
25	15511210043330985984000000
26	403291461126605635584000000
27	10888869450418352160768000000
28	304888344611713860501504000000
29	8841761993739701954543616000000
30	265252859812191058636308480000000
31	8222838654177922817725562880000000
32	263130836933693530167218012160000000
33	8683317618811886495518194401280000000
34	295232799039604140847618609643520000000
35	10333147966386144929666651337523200000000
36	371993326789901217467999448150835200000000
37	13763753091226345046315979581580902400000000
38	523022617466601111760007224100074291200000000
39	20397882081197443358640281739902897356800000000
40	815915283247897734345611269596115894272000000000
41	33452526613163807108170062053440751665152000000000
42	1405006117752879898543142606244511569936384000000000
43	60415263063373835637355132068513997507264512000000000

THOSE ARE THE FACTORIALS FROM 1 TO 43 .

```
10 REM FACTORIALS II
20 REM FRED PERILLO
30 DIM N(55), M(55)
40 LET I = K = 1
50 LET K6 = 6
70 MAT N = ZER
80 MAT M = ZER
100 N(0) = 55
105 M(0) = 55
110 FOR J2 = 1 TO 55
130 N(J2) = N(J2)* I + K
140 K = INP (N(J2)/10)
150 N(J2) = N(J2)-(K*10)
160 NEXT J2
170 IF N(55) > 0 THEN 450
180 J3 = 55
190 FOR K2 = 1 TO 55
200 M(K2) = N(J3)
210 J3 = J3-1
220 NEXT K2
230 FOR K4 = 1 TO 55
240 IF M(K4) < > 0 THEN 255
250 NEXT K4
255 FOR M9 = 1 TO K4-1
270 M(M9) = 5
280 NEXT M9
290 FOR M8 = K4 TO 55
300 M(M8) = M(M8) + 48
310 NEXT M8
320 CHANGE M TO N$
330 PRINT TAB(K6);I;N$
400 I = I + 1
410 IF I > 9 THEN 430
420 GO TO 110
430 K6 = 5
440 GO TO 110
450 I = I-1
460 PRINT 'THESE ARE THE FACTORIALS FROM 1 TO';I
999 END
```

ANALYSIS

The problem of truncation can be avoided by using base 1000 arithmetic.

The factorial is generated from 1 up until the partial product exceeds 1000. At that point the number is converted into its remainder modulo 1000. This is done in lines 410 and 420.

The 600 subroutine converts back to base 10 before the printing is done in line 620. The reader is invited to trace the program for N larger than 10 or so. That way he may gain some insight into the mechanism of the incrementation once base 1000 has been pulled into the algorithm.

The limit on this program is about 200!. It could be increased with the choice of a larger base.

A second program is also supplied which uses a slightly different algorithm and formats the output a bit differently.

59

Coin Flipper Simulator

Write a program to flip a coin any specified number of times.

Have the computer print out the actual result of each flip; that is, *H* when the coin comes up *heads* and *T* when it lands *tails*. A semicolon in your print statement will allow many entries per line.

Have the computer keep track of the tosses and print out after the run how many heads were tossed. Finally print out the ratio of heads to total tosses. It should be very close to .5 for a large number of tosses, if the program uses a fair coin.

References:

Steve Rogowski, **Computer Clippings,** p. 48.

```
HOW MANY FLIPS
? 45
H TH H TH H T TH T TH T T TH T TT TH T THH TH T THH T TT T T TH THHH TT
  18 HEADS IN  45 FLIPS!

HOW MANY FLIPS
? 100
TT T THHHH THH THHH THH T THH THH THH THH T TH TH TT T T TH TT THHH THH THHH T TH THH T TT TH TH T THH

HH THH TH T T THH THH THHHH THH TH T TH
  54 HEADS IN  100 FLIPS!

HOW MANY FLIPS
? 500
TH T THH T TH THHHHHHHH THHHHH THH T THH T TH T TH THHH TT TH THH TH TH TT TH T TH TT TH TT THH T THH T TH TT

HHHHH T TH TT TH TT THHH T THH THHHH TT TH TT TH THHHHHHHH TH T TH TTTH TH TH TT TH T THH T TH THH T

HH TH THH T THHH TT TT THHHHHH THHH T TH TH T TH TH TTTT TH THHH TT TH THHH T TH THHH T TH THHH T TH

HHHH TTTT TH T THHH TTT TH TTTTT THH TT T TH THH TT THHHH THHHHHHH T THH T THHH TH T TTTT THHH THH

TH T TH TT THH T THHH THH THHH TTT THHHH TTT TH TH T TH THH TTT THHHHH TH T TH T TH THH THHH THH T

HT TH THHH TH T TH T TH T TH TT THH TTT TH THHHHHHH THH THHH T THH TH TT TH THHH THH T THHH TH TH TH T TH

T THH T THHH TH THH TH TH TT THHHH TH TH TTTT TH TTH TH TH THHH TH TT THH TTT T TH THHH TH TH T
  256 HEADS IN  500 FLIPS!

HOW MANY FLIPS
? STOP
  PROGRAM STOPPED.
```

```
10 PRINT 'HOW MANY FLIPS'
20 INPUT N
30 LET C = 0
40 FOR X = 1 TO N
50 LET F = INT(2*RND)
60 IF F = 1 THEN 80
70 PRINT 'T';
75 GO TO 100
80 LET C = C + 1
90 PRINT 'H';
100 NEXT X
110 PRINT
120 PRINT C;' HEADS IN ';N;' FLIPS!'
125 PRINT
130 GO TO 10
200 END
```

ANALYSIS

This program is a useful simulation problem in elementary probability. It is valuable at most any grade level. Line 50 generates either a 0 or a 1. If 0 is generated, the program prints a *T* for tails and then flips again. If a 1 is generated it increments the heads counter in line 80 and prints an *H*.

When the limit set by the user is reached it prints the total heads and the total flips and then asks for another run.

The program could be expanded or just used as is to study the frequency and likelihood of repeated patterns.

60

The Old License Plate Trick

Suppose you agreed, as you were traveling along in your car, to write down the last two digits of the license plates of the next twenty cars to go by you.

One would have to agree that cars pass by completely at random in most situations, at least as far as the numbering on their plates in concerned. What do you think the chances are that in that group of twenty two-digit numbers, two of them will be the same? Remember there are 100 possible numbers from 00 to 99. You are going to record only the last two digits, not the whole plate number, and that only for the first twenty cars to go by.

Write a computer program to simulate the passage of this traffic. Then compute the probability that there will be a matching number in your list. As you generate license numbers at random, check them against others in your list until you have a match or until you've reached twenty cars. You can use two letters and four digits or three of each. It won't affect the computation of the probability. Generate random numbers and then use whatever BASIC function converts from internal code to character representation (CHR$, CHANGE, etc.).

Check your list for matches and compare these values to the computed values from your table. That is, verify that the chance for twenty cars producing a list of plates with a matching two-digit number is indeed almost 0.87.

References:

Fred Mosteller, **Probability and Statistics**.

Henry Adler, **Introduction to Probability and Statistics**, pp. 60–67.

CARS	PROBAB	CARS	PROBAB
1	0	16	.718402
2	9.99892E-03	17	.763456
3	2.97989E-02	18	.80367
4	5.89042E-02	19	.839008
5	9.65472E-02	20	.869597
6	.141716	21	.895676
7	.193209	22	.917584
8	.249684	23	.935716
9	.30971	24	.950501
10	.371838	25	.96238
11	.434651	26	.971785
12	.496838	27	.979121
13	.55722	28	.984758
14	.614777	29	.989026
15	.668712	30	.992209

```
10 PRINT 'CARS','PROBAB'
20 PRINT
30 FOR N=1 TO 30
40 LET Z=0
50 FOR X=1 TO N
60 LET T=LOG(100-(X-1))
70 LET Z=Z+T
80 NEXT X
90 LET R=N*LOG(100)
100 LET Y=EXP(Z-R)
110 PRINT N,1-Y
120 NEXT N
200 END
```

ANALYSIS

The output here is astounding to the amateur gambler. Only 13 cars are needed to make the outcome of the event more likely than not. With twenty cars the odds are a bettor's dream with almost 9 out of 10 tries producing the desired result.

The computation was done using the reasoning that the product of the probabilities for independent events will give the probability of all of them occuring. For the first car there is no chance that a match can occur. For the second car the probability is:

$$1 - (100/100) \cdot (99/100)$$

since there are only 99 numbers left that will produce no match. When the third car travels by only 98 numbers remain which will produce no match. Hence the probability of a match increases as the right hand term of the expression below decreases:

$$1 - (100/100) \cdot (99/100) \cdot (98/100)$$

The numerator is really $100-(x-1)$ where x is the number of cars. The denominator is 100 to the power x.

The computation was done using logarithms in lines 60 and 90. That way addition could be used and the incremented values would be more accurate than if partial products were accumulated. These are not logs to the base 10—although they would have worked fine—but rather natural logs (to the base e). Therefore the antilog and the subtraction of the logs to represent division are done in line 100 by raising the base e to the power of the log. It remains only to subtract the result from 1, since the probability which has been generated is that for *no match*. This is done in line 110.

ZJ	4914	WH	9253
WJ	2388	XG	3547
SF	7576	IH	8659
RK	2813	MP	1785
TY	8456	EL	1939
LG	6919	EG	0823
FZ	7979	YK	8806
KW	3626	FB	2358
PS	9091	QB	8216
IG	3207	HY	8586

```
100 RANDOMIZE
110 DIM A(50),A$(50)
120 LET A(0) = 7
130 FOR X = 1 TO 10
140 FOR X1 = 1 TO 2
150 LET A(1) = INT(25*RND) + 6
160 LET A(2) = INT(25*RND) + 6
170 LET A(3) = 5
180 FOR Y = 4 TO 7
190 LET A(Y) = INT(RND*10) + 48
200 NEXT Y
210 CHANGE A TO A$
220 PRINT A$,
230 NEXT X1
240 PRINT
250 NEXT X
260 END
```

ANALYSIS

This program simply generates a random list of twenty license plates. It generates random numbers between 65 and 90. These are the coded equivalents of the letters *A* through *Z*. Statements 150 and 160 generate the first two letters. Statement 170 assigns a *blank* to the third position in the plate field. The code for a blank on our system is 32, it may vary from system to system, as will the method for converting these numbers to their character representation. The Appendix lists the values for each character.

The loop from 180 to 200 generates numbers from 48 to 57 which are the coded equivalents of the digits 0 through 9. Four such digits are generated. The entire seven digit field specified in line 120 is converted into a string in loop 210. The actual printing is done by statement 224 which is still within the loop. The comma causes the cylinder to remain on that print line, so that two strings are printed ten times—loop from 130 to 250.

It has been left as an exercise for the reader to expand the program to not only generate the list but also to check it for a match in the last two places and then keep track of the results. See my book *Computer Clippings* for a more elaborate treatment of this problem. It is very similar to the *Birthday Problem* mentioned elsewhere in this volume.

A telephone number has seven digits. The first two are usually limited in a given area to a specified pair or two. Let's consider only the last five digits for this problem.

Compute the probability that in this five digit number, at least two of the digits match. Remember, any one of ten digits could fall in those five places. There are a total of 100,000 numbers which could occur. Compute how many permutations of ten digits will be needed to fill the five spots. Divide this by 100,000 to get the probability that no match has occured.

Write a computer program to generate five digit numbers at random. Devise a routine to test their digits for a match. If no match occurs call it a *Lola*. Print out the number of *Lolas* generated for a given set of five digit numbers. Compute what percentage of the total count were *Lolas*. Compare this with the computed probability. Which is more likely, a phone number with five distinct digits or one with at least a matching set of digits?

```
Zoerner L  17ConslRdClne -------869-6607
Zoll William L  BeverwyckLatham --785-5232
Zoller Albert J Jr 10CaroIneLathm 785-7089
Zoller Albert J Sr JohnPTaylorApts 274-4325
Zoller Jos J  SnydersLake ------283-1368
Zoltanski Joseph
            11DeerpathDrClne--869-3713
Zonitch J D  800-19thWvlt ------274-1420
Zonitch John  20BrentwdAv -----273-7242
Zonitch Mary Mrs  47CraigWvlt --273-5575
Zordan L J 6-9thWfd ----------237-5187
Zordan M J  12-9thWfd -------237-2541
Zorella A 2316-9thAvWvlt -------274-0455
Zorian Gregory T  397-4thAvNTy -235-5322
Zorian Mary A Miss
            1802-7thAvWvlt--271-8229
```

References:

Fred Mosteller, **Probability and Statistics**.

Henry Adler, **Introduction to Probability and Statistics**.

116

```
TELEPHONE BOOK PROBLEM!

HOW MANY TRIALS!
? 100
IN   100 TRIALS THERE WERE   25 LOLAS!
P(LOLA) =  .25

ANY MORE TO DO
? YES

HOW MANY TRIALS!
? 500
IN   500 TRIALS THERE WERE   137 LOLAS!
P(LOLA) =  .274

ANY MORE TO DO
? YES

HOW MANY TRIALS!
? 1000
IN   1000 TRIALS THERE WERE   315 LOLAS!
P(LOLA) =  .315

ANY MORE TO DO
? 5000
```

```
100 PRINT'TELEPHONE BOOK PROBLEM!'
110 PRINT
120 PRINT 'HOW MANY TRIALS!'
130 INPUT N
140 RANDOMIZE
150 LET S = 0
160 FOR W = 1 TO N
170 FOR X = 1 TO 5
180 LET Y(X) = INT(10*RND(-1))
190 NEXT X
200 FOR X = 1 TO 5
210 FOR B = X + 1 TO 5
220 IF Y(X) = Y(B) THEN 260
230 NEXT B
240 NEXT X
250 LET S = S + 1
260 NEXT W
270 PRINT 'IN ';N;' TRIALS THERE WERE ';S;' LOLAS!'
280 PRINT 'P(LOLA) = ';S/N
290 PRINT
300 PRINT 'ANY MORE TO DO'
310 INPUT A$
320 IF A$ = 'YES' THEN 110
330 END
```

ANALYSIS

The problem at hand is really how many numbers with five digits, and there are 100,000 of them—00000 to 99999—have digits which do not repeat.

The computation is equivalent to finding the number of permutations of ten digits into five places, i.e., ten things taken five at a time. That number is 30,240. The probability of no digit match is therefore, 30,240/100,000 or 0.3024.

In terms of the simulation, we are really generating telephone numbers from the last five digits in our random number generator in line 180. The loop in 200 through 240 checks for matching digits.

The crux of the problem is that there is almost a 70% chance of finding a five-digit telephone number in which, at least one number matches. The prefixes are avoided for obvious reasons. This can also be done with the four digits in a license plate number.

There are a myriad of options on this topic. Notice in the output that it took as many as 1000 trials to produce a reasonable value for the probability.

The area of a circle is πr^2. The area of a quadrant of a unit circle ($r=1$) is one-fourth of πr^2.

Suppose we place that quadrant within a square as shown below. The area of the square is 1, while the area of the quadrant is less than 1 and equal to $\frac{1}{4} \times \pi r^2$. If we were to generate points at random, so that every point fell within the confines of the square but sometimes within and sometimes without the quadrant of the circle, we would expect the frequency of each to be related to the ratio of their areas. If this random generation involved a large enough sample, the number of points falling within the quadrant as compared to the total generated would tend to equal the ratio of the area of the quadrant to the area of the square.

Derive an expression for that ratio. Certainly, it contains π. Program the computer to generate ordered pairs of coordinates at random. You will need to test whether the points are inside the quadrant. *Count **on** the circle as within it.* Vary the size of the sample. You should be able to approximate π with the value which you obtain.

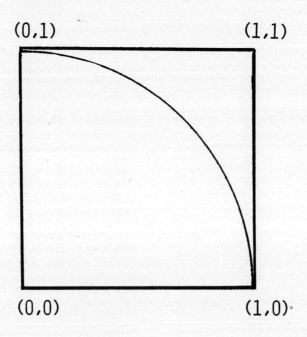

GENERATION OF PI BY RANDOM NUMBERS

```
HOW MANY TRIALS
? 100
FOR   100 TRIALS PI =   3.16



HOW MANY TRIALS
? 500
FOR   500 TRIALS PI =   3.240



HOW MANY TRIALS
? 1000
FOR   1000 TRIALS PI =   3.0640



HOW MANY TRIALS
? 5000
FOR   5000 TRIALS PI =   3.1520



HOW MANY TRIALS
? 10000
FOR   10000 TRIALS PI =   3.18240



HOW MANY TRIALS
? STOP
 PROGRAM STOPPED.
```

```
05 RANDOMIZE
10 PRINT 'GENERATION OF PI BY RANDOM NUMBERS'
20 PRINT
30 PRINT 'HOW MANY TRIALS'
40 INPUT N
45 LET A = B = 0
50 FOR Z = 1 TO N
60 LET X = RND(-1)
65 LET Y = RND(7)
70 IF X↑2 + Y↑2 > 1 THEN 90
75 LET A = A + 1
80 GO TO 110
90 LET B = B + 1
110 NEXT Z
115 PRINT A; ' SUCCESSES'
120 LET P = A/(A + B)
130 PRINT 'FOR ';N;' TRIALS PI = ';4*P
135 PRINT
140 PRINT
145 GO TO 20
200 END
```

ANALYSIS

Lines 60 and 65 generate the coordinates of a random point. The arguments of the RND function mean nothing on our computer system but on some systems they may be used to seed the starting point of the random number sequence.

Line 70 checks to see if the pair is within the quadrant of the circle. On the circle is considered inside for the purposes of this program. If the pair does fall within the sector, A is incremented by 1. The total number of points falling within or on the circle should be in the same ratio with the total points generated as the ratios of the areas of the two figures. This is $\pi/4$ to 1 since we chose a unit circle with area π. The ratio of the points should therefore be the same. Multiply the percentage of hits by 4 to get an approximation to π.

This multiplication is done in line 130 and the program is recycled. Notice the poor approximations which are produced. A large sample is required to produce even a small number of places for π. No wonder mathematicians have found this *dart tossing* technique inferior to the computation by arctangent series.

A series of parallel lines are ruled on a surface. The distance between the lines is equal. A needle, whose length is equal to the distance between the lines is dropped onto the surface. The probability that the needle crosses a line, rather than lies entirely within a space is $2/\pi$.

This has a number of interesting ramifications. It means that we can estimate π with a series of random occurrences. Dropping the needle a considerable number of times should produce say x crossings. This will mean that $x \cong 2/\pi$. So π is an approximation to $2/x$.

The horizontal position of the needle is immaterial. All lines are identical. The following diagram may help you set the problem up. It shows the result of a representative toss:

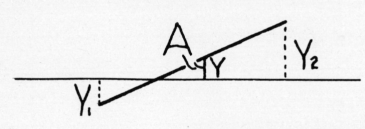

Note: Needle length is exaggerated for clarity.

A random number y is chosen between 0 and 1, since we took the spacing of the lines as one unit. Then A is chosen as a random angle between -90° and +90°, that is between $-\pi/2$ and $+\pi/2$ radians.

Y1 and *Y2* are determined by trigonometry. If the integer parts of *Y1* and *Y2* are different then the needle has crossed the line. If the integer parts are the same then the needle lies entirely within the space.

Write a program to simulate the dropping of a needle. Determine whether the needle crosses a line and keep track of the number of times it does. See how close an approximation to π this produces.

References:

Kemeny and Kurtz, **BASIC Programming**, p. 68.

Heinrich Dorrie, **100 Great Problems of Elementary Mathematics**, pp. 73-75.

```
HOW MANY TRIALS
? 100
THERE WERE  64 CROSSINGS IN  100 TRIALS!
PERCENT CROSSINGS WAS  63.999999 %
PI IS APPROX =  3.125

WANT TO TRY AGAIN
? YES
HOW MANY TRIALS
? 500
THERE WERE  314 CROSSINGS IN  500 TRIALS!
PERCENT CROSSINGS WAS  62.80 %
PI IS APPROX =  3.1847134

WANT TO TRY AGAIN
? YES
HOW MANY TRIALS
? 1000
THERE WERE  643 CROSSINGS IN  1000 TRIALS!
PERCENT CROSSINGS WAS  64.299999 %
PI IS APPROX =  3.1104199

WANT TO TRY AGAIN
? NOT REALLY
```

```
10 PRINT 'HOW MANY TRIALS'
20 INPUT N
30 RANDOMIZE
40 LET R = 57.295779
50 LET S = 0
60 FOR T = 1 TO N
70 LET Y = RND(-1)
80 LET X = RND(-2)
90 LET Z = SGN(X-.5)
100 LET A = Z*(90*RND(-3))/R
110 LET Y1 = Y-COS(A)/2
120 LET Y2 = Y+COS(A)/2
130 IF INT(Y1) = INT(Y2) THEN 150
140 LET S = S+1
150 NEXT T
160 LET F = S/N
170 PRINT 'THERE WERE ';S;' CROSSINGS IN ';N;' TRIALS!'
180 PRINT 'PERCENT CROSSINGS WAS ';100*F;' %'
190 PRINT 'PI IS APPROX = ';2/F
200 PRINT
210 PRINT 'WANT TO TRY AGAIN'
220 INPUT A$
230 IF A$ = 'YES' THEN 10
300 END
```

ANALYSIS

The first item to notice is that the value of π is quite unreliable even with a large number of trials. Only a large number of tosses can stabilize the value computed.

Line 40 in the program sets up the change of degrees to radians. Line 100 generates the angle between -90° and +90°. Lines 110 and 120 determine where the center of the needle falls relative to the space between the lines. Remember the needle is the same length as the spacing between the lines.

Line 130 tests for a crossing. Line 160 computes the crossing ratio and line 190 converts to π. The value of Y is the distance of the center of the needle from a line. In line 110 the cosine is divided by 2 because the needle has a length of 1 with half of it to the left of center. The right triangle relationships from the previous diagram help establish this more clearly.

121

The Matching Birthday Problem

What is the probability that in a group of thirty people at least two have the same birthday? If you trust your intuition on this problem one would probably guess that it was remote.

Most people are surprised to find that there is a 70% chance of two people having a matching birthday in a random group of 30 people.

Derive an expression to compute the probability when the size of the group varies. Have the computer produce a table showing the probabilities for from 1 to 40 people. The formula will involve the product of the probabilities for each successive person.

Once you've generated the table and confirmed the 70% figure, you can verify it experimentally. You could interrogate a large number of groups of people and then record your successes and failures. You should be successful 70% of the time for a large number of trials.

With the computer, however, we can use random numbers to simulate these events. Generate random numbers between 1 and 365. Let these represent a day of the year. In any group of 30 successive random numbers the probability that two of them are the same is almost 0.70. Once you've generated the 30 numbers test that group for a match. If successful increment the success parameter by 1. For a large number of simulations the number of successes divided by the total number of trials should just about equal the tabular probability for that number of people.

It's interesting to start out by letting the computer actually print out the random numbers and to test for a match by hand. You'll soon realize the value of having a routine in the program to do that for you.

Reference:

Mathematics, *Time-Life Books,* Chapter on *Probability*.

PEOPLE	PROBABILITY	PEOPLE	PROBABILITY
1	0	21	.443692
2	2.73824E-03	22	.475698
3	8.20267E-03	23	.507301
4	1.63512E-02	24	.538348
5	2.71312E-02	25	.568698
6	4.04592E-02	26	.598248
7	5.62241E-02	27	.626863
8	7.43271E-02	28	.65446
9	9.46153E-02	29	.680971
10	.116937	30	.706316
11	.141133	31	.730463
12	.167019	32	.753354
13	.194412	33	.774973
14	.223099	34	.795325
15	.252906	35	.81439
16	.283612	36	.832194
17	.31502	37	.848742
18	.346923	38	.864075
19	.379121	39	.87823
20	.41144	40	.89124

```
10 PRINT 'PEOPLE',
   'PROBABILITY'
30 PRINT
40 FOR N=1 TO 40
45 LET Z=0
50 FOR X=1 TO N
60 LET T=LGT(365-(X-1))
70 LET Z=Z+T
80 NEXT X
90 LET R=N*LGT(365)
95 LET Y=10↑(Z-R)
100 PRINT N,1-Y
110 NEXT N
200 END
```

292 114 241 210 186 362 312 342 157 136 212 253 126 281 202
123 148 109 156 21 278 283 293 236 128 106 162 214 192 117

152 144 267 46 213 123 328 105 185 249 328 98 111 275 135
229 303 269 23 198 92 175 346 31 137 199 282 216 20 263

185 327 95 365 180 364 263 313 133 149 1 341 81 210 176
236 41 130 17 202 9 364 234 167 280 295 244 61 52 286

99 93 3 135 148 106 125 113 198 140 183 280 227 155 313
188 168 7 37 342 164 209 279 334 151 224 238 142 36 180

325 64 41 11 333 40 168 153 340 327 55 69 73 336 57
218 173 210 333 269 7 353 137 360 340 206 108 333 325 111

242 105 331 255 61 64 259 247 254 66 82 352 118 245 105
69 323 172 180 39 351 157 39 149 314 189 284 125 153 123

116 138 192 229 273 196 67 299 291 234 129 136 242 300 21
177 340 38 125 93 347 200 351 14 359 238 309 269 332 165

117 267 240 22 161 183 24 68 210 28 157 15 169 240 175
69 182 268 227 285 337 355 61 275 310 254 253 168 49 5

105 330 18 351 97 270 24 312 185 148 105 355 350 48 224
262 89 170 24 184 273 155 200 94 320 244 21 90 22 322

22 336 330 133 92 241 309 63 190 106 104 123 111 185 310
244 228 243 20 131 104 5 355 257 240 96 84 355 357 252

```
02 RANDOMIZE
05 FOR M = 1 TO 10
10 FOR X = 1 TO 2
20 FOR Y = 1 TO 15
30 PRINT INT(365*RND) + 1;
40 NEXT Y
45 PRINT
50 NEXT X
54 PRINT
55 NEXT M
60 END
```

189   353   55   76   332   211   109   217   322   342   250   251   353   230   29
297   57   233   151   273   284   343   231   125   135   56   213   49   203   55

353 APPEARS TWICE!

324   359   337   251   297   257   326   11   57   243   217   210   38   163   274
174   45   160   196   100   301   175   171   186   308   174   100   133   262   13

174 APPEARS TWICE!

279   99   275   29   260   202   331   176   265   12   241   247   46   245   325
111   299   68   273   298   62   248   196   43   323   93   210   64   318   239

33   240   49   321   28   202   230   295   65   289   58   304   205   320   83
172   289   185   338   2   259   72   296   37   110   333   282   154   212   252

289 APPEARS TWICE!

333   101   167   92   152   82   223   239   157   259   137   324   344   249   221
189   236   83   203   111   299   69   275   303   76   288   317   43   324   97

324 APPEARS TWICE!

```
02 RANDOMIZE
04 DIM Z(100)
05 FOR M = 1 TO 10
07 LET I = 1
10 FOR X = 1 TO 2
20 FOR Y = 1 TO 15
30 LET Z(I) = INT(365*RND) + 1
35 PRINT Z(I);
37 LET I = I + 1
40 NEXT Y
45 PRINT
50 NEXT X
54 PRINT
60 FOR B = 1 TO 29
70 FOR A = B + 1 TO 30
80 IF Z(B) = Z(A) THEN PRINT Z(B);' APPEARS TWICE!' ELSE 170
170 NEXT A
180 NEXT B
190 PRINT
195 NEXT M
200 END
```

BIRTHDAY PROBLEM

HOW LARGE SHALL I MAKE THE SAMPLE
? 100
SUCCESSES   FAILURES
61   39

THE PROBABILITY THAT TWO BIRTHDAYS WILL MATCH
IN A GROUP OF 30 PEOPLE IS .61

BIRTHDAY PROBLEM

HOW LARGE SHALL I MAKE THE SAMPLE
? 200
SUCCESSES   FAILURES
142   58

THE PROBABILITY THAT TWO BIRTHDAYS WILL MATCH
IN A GROUP OF 30 PEOPLE IS .70999999

```
05 RANDOMIZE
10 PRINT 'ROGO'
20 PRINT 'BIRTHDAY PROBLEM'
30 PRINT
40 LET E = 0
50 LET C = 0
55 PRINT 'HOW LARGE SHALL I MAKE THE SAMPLE'
57 INPUT N
60 DIM N(30)
70 FOR D = 1 TO N
80 FOR X = 1 TO 30
90 LET N(X) = INT(365*RND(-1)) + 1
100 NEXT X
110 FOR B = 1 TO 29
120 FOR A = B + 1 TO 30
130 IF N(B) = N(A) THEN 170
140 NEXT A
150 NEXT B
155 LET E = E + 1
160 GO TO 180
170 LET C = C + 1
180 NEXT D
190 PRINT 'SUCCESSES','FAILURES'
200 PRINT C,E
210 PRINT
220 PRINT 'THE PROBABILITY THAT TWO
     BIRTHDAYS WILL MATCH'
230 PRINT 'IN A GROUP OF 30 PEOPLE IS ';C/N
240 PRINT
300 END
```

ANALYSIS

The first program prints out the probability table. It can be seen that it is more likely than not that there are at least two matching birthdays when the group of people is 23 or greater. A class with 30 students would have a 70% chance of having two matching birthdays.

The table is generated employing computations with logs to the base 10. Using the formula:

$$1 - \frac{365 \cdot 364 \cdot 363 \cdot 362 \cdot [365-(x-1)]}{365^x}$$

where x is the number of people in the group.

Each term is computed separately and the partial sum is accumulated in Z for the numerator. R is the log of the denominator. The subtraction and the anti-log are both done in line 95.

The second program confirms the computed probabilities. It assumes that the problem is equivalent to finding two matching numbers in a random group of numbers such that $1 < n < 365$. The program following the table generates 30 random numbers within those limits. The matches have been indicated. Out of 10 attempts, seven contain a match. Almost too good to be true. The reader is invited to search the lists where matches have not been indicated to see how time consuming a process it is.

The final program generates N sets of 30 numbers between 1 and 365. It then checks the list for a match and counts the number of successes.

The numbers are generated in the X loop at line 80. They are compared in nested B and A loops at 110 and 120. The variable C totals the successful matches, the variable E the failure to match. This searching process can be quite time consuming.

In conjunction with linear data regression one might want to know how closely two sets of measurements are related. One might, for example, be interested in how the I.Q. scores and the College Board Verbal Achievement scores correlate.

A correlation coefficient, which is always less than 1 is used. The closer the number is to 1, the stronger is the correlation. Correlation coefficients may also be negative indicating either a strong or weak inverse relationship. An example might be the number of degrees on the thermometer and the number of gallons of heating oil sold.

In order to compute the correlation coefficient one needs to know the following about the sample data:

n = the number of scores
$\overline{x}$ = the arithmetic mean of the first set
$\overline{y}$ = the arithmetic mean of the second set
σ_x = the standard deviation for the first set
σ_y = the standard deviation for the second set

Once these are known the coefficient can be computed using the following formula:

$$r = \frac{\sum(x_i - \overline{x})\ (y_i - \overline{y})}{n\sigma_x\sigma_y}$$

where r is the coefficient of correlation.

References:

P. Calingaert, **Principles of Computation**, pp. 78–80.

A set of scores upon which you may wish to do the correlation appears in the appendix.

```
HOW MANY DATA PAIRS? 50
PAIRED SCORES****
```

127	591
117	446
127	538
120	433
125	696
125	591
133	506
1U9	499
122	519
134	650
123	525
125	387
122	519
120	446
132	519
134	492
126	637
141	598
141	545
114	420
110	486
141	486
128	611
118	453
123	453
128	644
121	571
121	578
122	433
125	578
126	552
132	630
132	552
134	708
127	506
121	512
138	650
125	499
132	519
108	511
121	446
128	677
126	538
132	532
123	617
126	644
125	584
122	479
120	519
133	571

```
100 PRINT 'HOW MANY DATA PAIRS';
110 INPUT N
120 PRINT 'PAIRED SCORES****'
130 LET S1 = S2 = S3 = S4 = S5 = 0
140 PRINT
150 FOR X = 1 TO N
160 READ A,B
170 PRINT A,B
180 LET S1 = S1 + A
190 LET S2 = S2 + B
200 LET S3 = S3 + (A*B)
210 LET S4 = S4 + (A*A)
220 LET S5 = S5 + (B*B)
230 NEXT X
240 PRINT
250 PRINT 'SUM X = ';S1,'SUM Y = ';S2
260 PRINT 'SUM X*Y = ';S3
270 PRINT 'SUM X12 = ';S4
280 PRINT 'SUM Y12 = ';S5
290 PRINT
300 LET C = N*S3-S1*S2
310 LET D = SQR((N*S4-S112)*(N*S5-S212))
320 PRINT 'CORRELATION COEFFICIENT = ';C/D
330 PRINT
340 DATA 127,591,117,446,127,538,120,433,125,696,125,591,133,506
350 DATA 119,499,122,519,134,650,123,525,125,387,122,519,120,446
360 DATA 132,519,134,492,126,637,141,598,141,545,114,420,110,486
370 DATA 141,486,128,611,118,453,123,453,128,644,121,571,121,578
380 DATA 122,433,125,578,126,552,132,630,132,552,134,708,127,506
390 DATA 121,512,138,650,125,499,132,519,108,511,121,446,128,677
400 DATA 126,538,132,532,123,617,126,644,125,584,122,479,120,519
410 DATA 133,571
420 END
```

ANALYSIS

The paired scores used here represent I.Q. scores and SAT verbal scores for a group of fifty students. They are the same data pairs that appear in the Appendix.

The loop from 150 to 230 does most of the work. In this loop five partial sums are incremented. Each separate score, each cross-pair, and each score squared are incremented. Their sums are provided as output in statements 250 through 280.

The *Pearson-Product Moment* formula is used. The sums generated are used in lines 300 through 320 to compute the coefficient. There are less efficient algorithms which involve more computation but which do give more insight into the nature of the correlation process. These are easily programmed.

A dispersion graph is also a possible extension of this program.

```
SUM X =    6295   SUM Y =    27096
SUM X*Y =     3422693
SUM X12 =    795067
SUM Y12 =    14968956

CORRELATION COEFFICIENT =   .42128899
```

Write a program to find the arithmetic, geometric and harmonic means for a set of scores. You may use the scores in the appendix or your own set. Compare each measure of central tendency and try to establish a relationship between them. You may use the formula sheet in the appendix to establish the correct mathematical procedure for computing each of these means.

Find the mean deviation from the mean. Compute a variance and a standard deviation.

There is a best way in which to compute each of the required quantities. The order in which you establish variables is important. For example, it is best to sum up the squares of the deviations in the same loop as the generation of those squares. Obviously you cannot compute deviations from the mean in the same loop that is used to read the scores. The mean is not available until all scores have been summed.

Do not generate a table of intermediate values. Have the computer sort the scores, then find the median and the mode. Print out the list of sorted scores.

```
ARITHMETIC MEAN =   75
GEOMETRIC MEAN =  74.3258
HARMONIC MEAN =  73.6484

MEAN DEVIATION FROM MEAN =  8.57143
VARIANCE IS  100
STANDARD DEVIATION =  10
```

```
0005 DIM A(50),B(50),H(50)
0100 PRINT 'LINEAR DATA REGRESSION'
0110 PRINT
0120 PRINT 'HOW MANY PIECES OF DATA'
0130 INPUT N
0140 LET G=J=C=H=T=0
0150 PRINT
0160 FOR D=1 TO N
0170 READ A(D)
0180 LET C=C+A(D)
0190 LET H=H+(1/A(D))
0200 LET W=LOG(A(D))
0210 LET T=T+W
0220 NEXT D
0230 LET E=C/N
0240 LET T=T/N
0250 LET T=EXP(T)
0260 LET K=N/H
0270 PRINT 'ARITH METIC MEAN = ';E
0280 PRINT 'GEOMETRIC MEAN = ';T
0290 PRINT 'HARMONIC MEAN = ';K
0300 PRINT
0310 FOR F=1 TO N
0320 LET B(F)=ABS(A(F)-E)
0330 LET G=G+B(F)
0340 LET H(F)=B(F)↑2
0350 LET J=J+H(F)
0360 NEXT F
0370 LET M=J/N
0380 PRINT 'MEAN DEVIATION FROM MEAN = ';G/N
0390 PRINT 'VARIANCE IS ';M
0400 PRINT 'STANDARD DEVIATION = ';SQR(M)
0410 PRINT
0420 DATA 30,32,40,27,29
0430 END
```

ANALYSIS

There is no need to sort the data for the first part of this problem. The program presented does all the incrementing and reading of scores in the D loop from lines 160 to 220.

The F loop beginning in 310 computes the deviations and squares of the deviations. There was really no need to subscript these variables. That was done in anticipation of future additions and possibly a more detailed printout.

The mode, median, root-mean-square and coefficient of variation are left as an exercise. The formulas for these measures can be found in the table in the Appendix.

Roots of Complex Numbers

Write a program to take the coefficients of a complex number in rectangular form: $a + bi$ and convert it to polar form: $k(\cos \theta + i \sin \theta)$.

The program should receive as input for this part only the variables a and b. Compute the modulus k and the angle θ within your program.

The second part of the program should take any root of the complex number. Input here should only be what root is to be taken as an integer. Compute the roots with the numbers in polar form, then convert back to rectangular and printout the results. Have the computer print all the roots. There will be n complex nth roots in rectangular form.

There are, for example three cube roots of the number 8. One of them is 2. What are the other two?

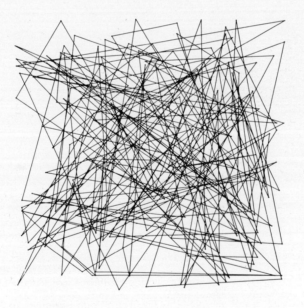

Reference:

Julian Mancill, **Modern Analytical Trigonometry**, pp. 167–185.

```
THIS PROGRAM TAKES THE ROOTS OF A COMPLEX NUMBER!
INPUT THE REAL AND IMAGINARY COEFFICIENTS.
? 8,0


  8 + 0 I  IN POLAR FORM IS
  8 * ( COS 0  + I SIN 0 )


WHICH ROOT DO YOU WISH
? 3

THE  3 ROOTS ARE:


  2 + 0 I
 -1 + 1.73205 I
 -.999999 + -1.73205 I




THIS PROGRAM TAKES THE ROOTS OF A COMPLEX NUMBER!
INPUT THE REAL AND IMAGINARY COEFFICIENTS.
? 16,0


  16 + 0 I  IN POLAR FORM IS
  16 * ( COS 0  + I SIN 0 )


WHICH ROOT DO YOU WISH
? 4

THE  4 ROOTS ARE:


  2 + 0 I
  0 + 2 I
 -2 + 0 I
  0 + -2 I
```

```
100 PRINT 'THIS PROGRAM TAKES THE ROOTS
    OF A COMPLEX NUMBER!'
110 PRINT 'INPUT THE REAL AND IMAGINARY
    COEFFICIENTS'
120 INPUT A,B
130 PRINT
140 LET P = 3.1415927
150 LET R1 = 57.29577
160 LET R = SQR(A↑2 + B↑2)
170 LET T = ATN(B/A)
180 PRINT A;' + ';B;' I  IN POLAR FORM IS'
190 PRINT R;'* (COS ';T*R1;' + I SIN ';T*R1;')'
200 PRINT
210 PRINT 'WHICH ROOT DO YOU WISH'
220 INPUT N
230 PRINT
240 PRINT 'THE ';N;' ROOTS ARE:'
250 PRINT
260 FOR K = 0 TO N-1
270 PRINT R↑(1/N)*(COS((T + (2*K*P))/N));
280 PRINT ' + ';R↑(1/N)*(SIN((T + (2*K*P))/N));' I'
290 NEXT K
300 END
```

ANALYSIS

This program will take a number in rectangular form and convert it to polar form. This is done in line 190. The relationship:

$$R \cdot (\cos \theta + i \sin \theta)$$

is used where: $R = \sqrt{a^2 + b^2}$ and $\tan \theta = b/a$ from the complex form $a + bi$.

The n roots are taken in statements 260 through 290. The formula is given in the loop. The variable P is the value of π and is defined in line 140. The root to be taken is stored in N and K goes from 0 to $N-1$ as it computes each root.

Notice the truncation that takes place in the output due to the use of the power series for computing trig functions. The numbers whose exponents of 10 are negative and large are approximations to zero. The four roots of 16 should therefore be read as 2, -2, 2i and -2i. It is left as an exercise for the reader to remedy this problem.

Suppose it is necessary to find the area under a curve between two vertical lines and the *x*-axis. The calculus tells us it can be done by integration. This is a challenging yet useful problem for a computer. It is difficult to program the computer to do integration by algebraic techniques, such as the manipulation of exponents, or by parts. It is not impossible. A clever way around the problem is as follows.

Imagine the *x*-axis to be the side of a rectangle, with the lines $x = a$ and $x = b$ forming the pair of sides perpendicular to the *x*-axis. It remains only to place the side parallel to the *x*-axis to complete the rectangle. That side should be placed so that it includes the entire curve in question.

If one were to generate random numbers between the limits imposed by the length and width of the rectangle, and if these random numbers were used as coordinates for points, then either of two possibilities exist. All points would fall within the rectangle but some would be above the curve and others below it. The number of points falling below the curve would be in direct proportion to the area beneath the curve. If one were to generate a sufficient number of points, the total number of points falling beneath the curve or on the curve divided by the total number of points generated should be in the same proportion as the area beneath the curve is to the total area of the rectangle. Since we can compute the area of the rectangle it should be possible to compute the area under the curve.

Write a program to calculate this area. You will need to input the limits of integration *a* and *b*, the equation of the curve, as well as the height of the rectangle. It is convenient to make the height of the curve the maximum value of the function at either point *a* or *b*. A more sophisticated program could compute this height from the values of the function at those points.

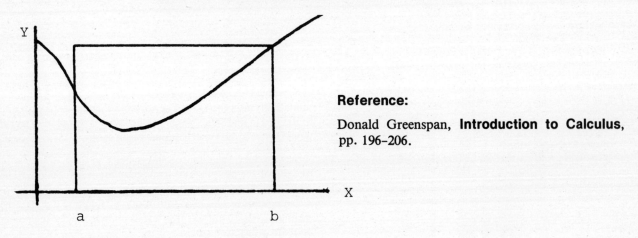

Reference:

Donald Greenspan, **Introduction to Calculus**, pp. 196–206.

```
INPUT LIMITS? 0,2.5
HOW MANY POINTS? 500
AREA OF RECTANGLE IS  2.7494
% AREA UNDER CURVE IS  66.2  %
AREA UNDER F(X) =  1.82011
```

```
110 PRINT "INPUT LIMITS";
120 INPUT A,B
130 PRINT "HOW MANY POINTS";
140 INPUT N
142 LET C=A
143 FOR X=A TO B STEP (B-A)/50
144 IF SIN(X)<SIN(C) THEN 146
145 LET C=X
146 NEXT X
147 LET C=1.1*SIN(C)
150 LET S=0
160 LET R=(B-A)*C
170 RANDOMIZE
180 PRINT "AREA OF RECTANGLE IS ";R
190 FOR W=1 TO N
200 LET X=((B-A)*RND(X))+A
210 LET Y=C*RND(X)
220 IF Y>SIN(X) THEN 240
230 LET S=S+1
240 NEXT W
250 LET R1=(S/N)*R
255 PRINT "% AREA UNDER CURVE IS ";(100*S)/N; "  %"
260 PRINT "AREA UNDER F(X) = ";R1
270 END
```

ANALYSIS

The problem is short but useful in showing the variability but relative consistency of random numbers.

Because the nature of the function cannot be predicted in advance, it is necessary to determine the maximum value of the function over the range a to b, so the rectangle will enclose the functional area. A more sophisticated program would search for maxima between a and b for they may not always occur at the limits of integration. Lines 142 through 146 search the function (a sine curve in this example) for a maximum and then add 10% for insurance in line 147. The rectangle may be larger than the curve itself is high.

The sine function is used here. Line 160 computes the area of the rectangle. The loop beginning in 190 generates n random pairs (x,y) between the limits a and b. Statement 220 tests the pairs to see if they fall beneath the curve. Statement 230 keeps track of successes. Line 170 insures a unique run each time the program is run. We don't want these darts to land in the same holes every time!

Notice the technique for generating random numbers within a certain interval. We multiply the random number which will always start out as an eight digit decimal between 0 and 1, by the difference between the end-points. We then add to this the lower limit.

133

Compute the area under a curve included between two vertical lines which represent the limits of integration. The area will be bounded by the curve, the vertical lines and the *x*-axis.

Break the area under the curve down into small trapezoids that are equal in height *(their height runs along the x-axis)* but have varying bases. Compute the area of each and sum them to approximate the area under the curve. Introduce the function into the program by means of the DEF FNA(X) statement.

As input you should include the limits of integration, the number of trapezoids, and the function to be integrated. Start with a small number of trapezoids and you will observe that the approximation, in most cases, becomes more accurate as the number of trapezoids increases.

There is no need to avoid functions whose upper limit is ∞. These can be taken care of by using a large number as an approximation to infinity.

If you've had any calculus, you might wish to integrate the function by traditional methods and compare the two values. It is a real challenge to get the computer to integrate algebraically but it can be done.

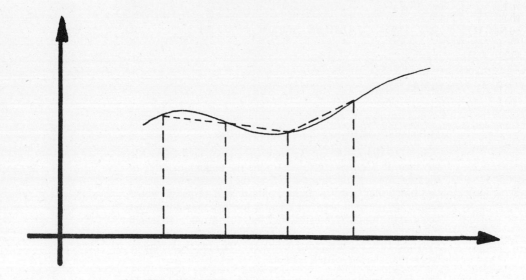

Reference:

Most any Calculus text will have a treatment of the trapezoidal rule.

Donald Greenspan, **Introduction to Calculus**, pp. 196–206.

```
THE LIMITS OF THE AREA AND NO OF TRAPEZOIDS PLEASE
? 0, 6, 500
AREA UNDER THE CURVE BETWEEN   0 AND   6 IS
 197.99978
DO YOU WISH TO RUN AGAIN (Y OR N)
? Y

THE LIMITS OF THE AREA AND NO OF TRAPEZOIDS PLEASE
? 0, 6, 1000
AREA UNDER THE CURVE BETWEEN   0 AND   6 IS
 197.99928
DO YOU WISH TO RUN AGAIN (Y OR N)
? N
```

```
10 DEF FNA(X) = X12 + 6*X + 3
15 PRINT
20 PRINT 'THE LIMITS OF THE AREA AND NO OF TRAPEZOIDS PLEASE'
30 INPUT A,B,N
40 LET H = (B-A)/N
50 LET S = 0
60 FOR X = A TO B-H STEP H
70 LET R = (FNA(X) + FNA(X + H))*(H/2)
80 LET S = S + R
90 NEXT X
100 PRINT 'AREA UNDER THE CURVE BETWEEN ';A;' AND ';B;' IS'
110 PRINT S
120 PRINT 'DO YOU WISH TO RUN AGAIN (Y OR N)'
130 INPUT A$
140 IF A$ = 'Y' THEN 15
200 END
```

ANALYSIS

The formula for the trapezoidal rule is familiar to every calculus teacher.

Line 40 computes the height of each trapezoid. Line 70 computes the lengths of the bases. Study the design of the incrementers for each loop. They are placed so as to include the endpoints of the area only once each. All intermediate bases are used twice.

Line 80 sums up the areas of each trapezoid. The function may be changed by simply inserting a new one in line 10.

Einstein's Energy Equation

Write a computer program to print out the amount of energy available from a given unit of mass and its cost.

Use Einstein's famous formula: $E = mc^2$. Input the mass in kilograms. The constant c will have to be in the proper units. Have the energy printed out in *joules* and *kilowatt hours*. Assume that a kilowatt hour cost 1¢ *(would that it were ever so!)*.

The table should go from 10 to 100 kilos. The conversion factors and speed of light in the proper units should be available in any elementary physics text.

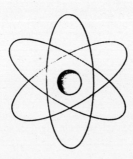

JOULES	KW-HOURS	DOLLARS	MASS (KG)
9.00000E+17	2.50000E+11	2.50000E+09	10
1.80000E+18	5.00000E+11	5.00000E+09	20
2.70000E+18	7.50000E+11	7.50000E+09	30
3.60000E+18	1.00000E+12	1.00000E+10	40
4.50000E+18	1.25000E+12	1.25000E+10	50
5.40000E+18	1.50000E+12	1.50000E+10	60
6.30000E+18	1.75000E+12	1.75000E+10	70
7.20000E+18	2.00000E+12	2.00000E+10	80
8.10000E+18	2.25000E+12	2.25000E+10	90
9.00000E+18	2.50000E+12	2.50000E+10	100

```
05 PRINT '';
10 PRINT 'JOULES','KW-HOURS','DOLLARS','MASS (KG)'
20 PRINT
30 LET C=3E8
40 FOR M=10 TO 100 STEP 10
50 LET E=M*C12
55 LET D=E/3.6E8
60 PRINT E,E/3.6E6,D,M
70 NEXT M
100 END
```

ANALYSIS

This table is an astounding illustration of the energy available to us from the direct conversion of matter to energy. Einstein's formula is used in line 50. The speed of light is taken as 300,000,000 meters per second. The E in line 30 indicates exponential notation and can be thought of as replacing *times 10 raised to the power*.

Statement 60 converts from joules to kilowatt-hours and also converts the cost to dollars.

Centigrade to Farenheit and Back

Write a program to print a table for converting from farenheit temperature to centigrade and vice-versa.

Let the units on the left be integers. use the entire print line for the table. The print-outs should be side-by-side. The formulas for the conversion are well known, and you should have no trouble finding them. Have the table go between whatever limits you wish.

Label all columns to avoid confusion. You might also compute the temperature in degrees Kelvin and Rankin if you wish to expand your program.

FAHRENHEIT	CENTIGRADE	CENTIGRADE	FAHRENHEIT
0	-17.7778	0	32
5	-15	5	41
10	-12.2222	10	50
15	-9.44444	15	59
20	-6.66667	20	68
25	-3.88889	25	77
30	-1.11111	30	86
35	1.66667	35	95
40	4.44444	40	104
45	7.22222	45	113
50	10	50	122
55	12.7778	55	131
60	15.5556	60	140
65	18.3333	65	149
70	21.1111	70	158
75	23.8889	75	167
80	26.6667	80	176
85	29.4444	85	185
90	32.2222	90	194
95	35	95	203
100	37.7778	100	212
105	40.5556	105	221
110	43.3333	110	230
115	46.1111	115	239
120	48.8889	120	248
125	51.6667	125	257
130	54.4444	130	266
135	57.2222	135	275
140	60	140	284
145	62.7778	145	293
150	65.5556	150	302
155	68.3333	155	311
160	71.1111	160	320
165	73.8889	165	329
170	76.6667	170	338
175	79.4444	175	347
180	82.2222	180	356
185	85	185	365
190	87.7778	190	374
195	90.5556	195	383
200	93.3333	200	392
205	96.1111	205	401
210	98.8889	210	410

```
10 PRINT 'FAREN','CENT','CENT','FAREN'
20 FOR F = 0 TO 212
30 PRINT F,(5/9)*(F-32),F,(9/5)*F + 32
40 NEXT F
50 END
```

Einstein predicted that as a body's speed increased toward the velocity of light its mass would increase without limit according to the relationship given below.

He used this to predict the fact that nothing could exceed the speed of light. He reasoned that at the speed of light the mass of the moving body would become infinite and would then require an infinite force to keep it in motion. Since no infinite forces exist, there could be no doubt that nothing with a mass can move at the velocity of light.

Some very tiny particles move very close to the speed of light. Write a program to compute for a given mass just how much of an increase would take place as the velocity of the body increased up to *c*—the velocity of light. Let your data be output in tabular form showing the velocity as a scalar and also as a function of its ratio to *c*.

Look up the velocity of light in any system, then keep *v* in the same units. Mass units will be identical to the units you input for rest mass.

A similar formula exists for relative time dilation, and for linear contraction. These are known as Lorentz' Transformations. You may wish to investigate what happens to time as a body moves toward the speed of light.

v = velocity of the body
c = velocity of light
m_o = rest mass
m_r = relative mass

$$m_r = m_o \sqrt{1/[1 - (v^2/c^2)]}$$

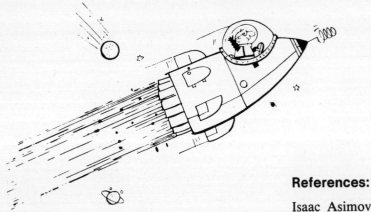

References:

Isaac Asimov, **Understanding Physics, Volume I, Force, Motion and Heat.**

```
INPUT THE REST MASS OF THE BODY IN KILOS:
? 85
VELOCITY OF LIGHT IS 2.99793E+08 METERS/SEC!

RELATIVE MASSVELOCITY(MPS) % OF C

85              0               0 %
85.0005         1.00000E+06     .333564 %
85.0019         2.00000E+06     .667128 %
85.0043         3.00000E+06     1.00069 %
85.0076         4.00000E+06     1.33426 %
85.0118         5.00000E+06     1.66782 %
85.0171         6.00000E+06     2.00138 %
85.0232         7.00000E+06     2.33495 %
85.0303         8.00000E+06     2.66851 %
85.0383         9.00000E+06     3.00207 %
85.0473         1.00000E+07     3.33564 %
85.0573         1.10000E+07     3.6692 %
85.0682         1.20000E+07     4.00277 %
85.08           1.30000E+07     4.33633 %
85.
STOP AT LINE 230
```

```
100 PRINT 'INPUT THE REST MASS OF THE BODY IN KILOS'
110 INPUT M
120 GO TO 170
130 PRINT 'RELATIVE MASS','VELOCITY(MPS)','% OF C'
140 PRINT
150 GO TO 210
160 PRINT
170 LET C=2.997925*10↑8
180 PRINT 'VELOCITY OF LIGHT IS ';C;' METERS/SEC!'
190 PRINT
200 GO TO 130
210 FOR V=0 TO 3E8 STEP 100000
220 LET R=M*SQR(1/(1-(V↑2/C↑2)))
230 PRINT R,V,(100*V)/C;'%'
240 NEXT V
250 END
```

ANALYSIS

This program illustrates the tremendous speed which must be attained to experience increases in relative mass as predicted by the Lorentz Transformation.

Even at 4 million meters per second a body with a mass of 85 kilograms at rest would have a mass of only 85.007 kilos. Measurable certainly, but hardly significant.

In the program line 170 defines the speed of light in meters per second. Statement 220 computes the relative mass while line 230 computes and prints the percentage of c that has been attained.

The following physics problem may seem to defy solution.

The diagram below is that of an infinite network of equal resistors. The difference of potential across the circuit is irrelevant. The idea is to find the total resistance of the network.

Remember resistances in series simply add up as they are, and resistances in parallel add up as reciprocals. Be sure to get the actual formulas exact.

THE CIRCUIT

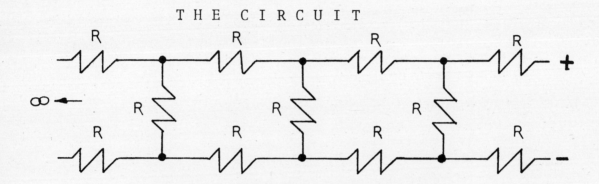

HINT: Careful study will lead you to the conclusion that a continued fraction will aid in the solution. Review what you know about them and let the computer evaluate the one you come up with. The answer involves the square root of three. Leave your answer in those terms.

References:

Kemeny and Kurtz, **Basic Programming**, pp. 124–126.

Sears and Zemansky, **University Physics**.

A. Ya. Khinchin, **Continued Fractions**.

FOR THE NETWORK BELOW, THE TABLE GIVES TOTAL
RESISTANCE

```
--*****-------****-------****-------

      *            *            *
      *            *            *
      *            *            *

--*****-------****-------****-------
```

NUMBER OF TERMS	PARTIAL SUM
1	2.75
2	2.7333333
3	2.7321428
4	2.7320574
5	2.7320513
6	2.7320508
7	2.7320508
8	2.7320508

```
10 PRINT 'FOR THE NETWORK BELOW, THE TABLE GIVES TOTAL'
15 PRINT '            RESISTANCE'
20 PRINT
21 PRINT '    --****-----****-----****-----'
22 PRINT '    R  1    1    1'
23 PRINT '       *    *    *'
24 PRINT '       *    *    *'
25 PRINT '       *    *    *'
26 PRINT '       1    1    1'
27 PRINT '    --****-----****-----****-----'
28 PRINT '    R    R    R'
29 PRINT
35 PRINT 'NUMBER OF TERMS',,'PARTIAL SUM'
40 PRINT
102 LET X = 3
105 FOR N = 1 TO 8
115 LET V = ((3*X) + 2)/(X + 1)
120 LET M = V
125 PRINT,N,,M
135 LET X = M
140 NEXT N
200 END
```

ANALYSIS

The program is based on a continued fraction. The actual value to which the resistance converges is $1 + \sqrt{3}$.

The general equation is:

$$R_n = 2R + 1/[(1/R + (1/R_{n-1})]$$

where $R_1 = 2R + R$ and contains:

$$R_2 = 2R + 1/[(1/R) + (1/R_1)] \text{ and contains:}$$

.

.

.

.

Line 115 computes the continued fraction and line 135 saves the terms computed in the previous run for inclusion in the subsequent run.

141

Write a program that will ask a young lover a few questions, and then write a letter to his girl friend on the basis of his answers.

INPUT statements of the form: 20 INPUT A$ are called for here. The remainder of the text is a problem for your own imagination.

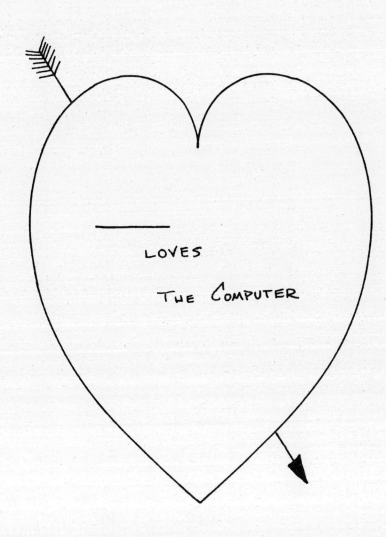

```
WHAT IS YØUR FIRST NAME AND HEIGHT (IN INCHES)? STEVE,68
WHAT IS YØUR GIRLFRIENDS FIRST NAME AND HEIGHT (IN INCHES)? JEANNETTE,60
WHAT CØLØR HAIR AND EYES DØES SHE HAVE?  BRØWN,HAZEL
HER BEST FEATURE IS HER (WHAT)? SMILE
WHEN DID YØU MEET AND WHERE? 1952,THIRD GRADE
WHAT IS ØNE THING YØU LIKE TØ DØ TØGETHER
? TRAVEL
WHEN DID YØU SEE HER LAST? THIS MØRNING

MY DEAREST JEANNETTE,
        I HAVE MISSED YØU SØ SINCE THIS MØRNING
I LØVE THE TØUCH ØF YØUR BRØWN HAIR
I LØVE TØ GAZE DØWN INTØ YØUR BEAUTIFUL HAZEL EYES
BUT MØST ØF ALL I LØVE YØUR SMILE
MY LUCKIEST DAY WAS WHEN I MET YØU AT THIRD GRADE
DØ YØU REMEMBER-IT WAS 1952
TØNIGHT LETS GET TØGETHER AND TRAVEL
UNTIL THEN-ALL MY LØVE
            STEVE
```

```
10 PRINT 'WHAT IS YOUR FIRST NAME AND HEIGHT (IN INCHES)';
20 INPUT X1$,X2
30 PRINT 'WHAT IS YOUR GIRLFRIENDS FIRST NAME AND HEIGHT (IN INCHES)';
40 INPUT X3$,X4
50 PRINT'WHAT COLOR HAIR AND EYES DOES SHE HAVE';
60 INPUT X5$,X6$
70 PRINT'HER BEST FEATURE IS HER (WHAT)';
80 INPUT Y3$
90 PRINT'WHEN DID YOU MEET AND WHERE';
100 INPUT X7$,X8$
110 PRINT'WHAT IS ONE THING YOU LIKE TO DO TOGETHER'
120 INPUT X9$
130 PRINT'WHEN DID YOU SEE HER LAST';
140 INPUT Y2$
150 IF X2> =X4 THEN Y1$ ='DOWN' ELSE Y1$ ='UP'
155 PRINT
160 PRINT 'MY DEAREST ';X3$;','
170 PRINT TAB(5);'I HAVE MISSED YOU SO SINCE ';Y2$
180 PRINT 'I LOVE THE TOUCH OF YOUR ';X5$;' HAIR'
190 PRINT'I LOVE TO GAZE ';Y1$;' INTO YOUR BEAUTIFUL ';X6$;' EYES'
200 PRINT 'BUT MOST OF ALL I LOVE YOUR ';Y3$
210 PRINT 'MY LUCKIEST DAY WAS WHEN I MET YOU AT '; X8$
220 PRINT'DO YOU REMEMBER-IT WAS ';X7$
230 PRINT 'TONIGHT LETS GET TOGETHER AND ';X9$
240 PRINT 'UNTIL THEN-ALL MY LOVE'
250 PRINT TAB(10);X1$
260 END
```

75

Numeric Sorter Routine
(Alphanumeric Option)

Write a program to sort a list of numbers entered at random.

The numbers may be any real numbers including zero and the negative reals. This is a required subroutine for most any statistical analysis. Remember numbers may appear more than once so take that into account.

This is an excellent exercise in using subscripted variables. It is a challenging program. Be sure to work the flow chart for this one out carefully. It will save considerable time and effort in the programming phase of the problem.

Do not use an INPUT statement here. It is quite time-consuming when used with extended data manipulation.

Include all data in a DATA statement at the end of the program.

A simple addition or modification to this program allows for the sorting of alphanumeric data (letters) which is a means of putting names, etc., in alphanumeric order.

```
16 PIECES TO BE SORTED
NUMERIC SORTER***


SORTED LIST****
  89  56  45  35  34  34  13  12  12  7  6  5  4  3  2  1
```

```
05 DIMA(100)
6 PRINT 'HOW MANY PIECES TO BE SORTED'
8 INPUT N
10 PRINT 'NUMERIC SORTER***'
20 PRINT
30 FOR X = 1 TO N
40 READ A(X)
45 NEXT X
50 PRINT
55 FOR K = N TO 2 STEP -1
60 LET G = A(K)
70 FOR J = 1 TO K-1
80 IF G > = A(J) THEN 200
90 LET T = G
100 LET G = A(J)
110 LET A(J) = T
200 NEXT J
210 LET A(K) = G
220 NEXT K
230 PRINT 'SORTED LIST ****'
240 FOR B = N TO 1 STEP -1
250 PRINT A(B);
260 NEXT B
300 DATA 12,34,35,45,56,12,13,1,2,3,4,5,6,7,89,34
400 END
```

ANALYSIS

Many studies have been done involving the efficiency fo sorting algorithms. This program is by no means the most economical way to sort numbers. It is, however, relatively simple to understand.

The X loop in 30 reads the data. The K loop takes the last number in the list and compares it with the other N-1 numbers which remain. The number in hand is G. When G is bigger than an item in the list, we go on to the next item. When G is smaller we put G back in the list, pick up the other number (line 100) call *it* G and then continue. When we finish one pass through the J loop we have found the biggest number. It is stored for future printing in line 210. It could actually be printed right there. The computer now repeats the process with the list one item smaller.

Notice the use of T in 90 and 110. T serves as a place to put G until we can pick up $A(J)$. When we tag $A(J)$ as the new G, we then go to T to pick up what's there (the old G) and put it back in the list as $A(J)$.

144

17 STRINGS TO BE SORTED

ALPHAMERIC SORTER*****

ALPHABETIZED LIST.....
ADAMS BAILEY BENCH COHEN CONWAY IKE JONES LEO LONG
LONGDO MEEMOM PEEPOP ROGO SHEAR SMITH WHALEN WHEELER

```
  5 DIM A$(5Ø)
 1Ø READ N
 2Ø PRINT N;"STRINGS TO BE SORTED"
 3Ø PRINT
 35 PRINT "ALPHAMERIC SORTER*****"
 4Ø PRINT
 5Ø FOR X=1 TO N
 6Ø READ A$(X)
 7Ø NEXT X
 8Ø PRINT
 9Ø FOR K=N TO 2 STEP -1
1ØØ LET G$=A$(K)
11Ø FOR J=1 TO K-1
12Ø IF G$>=A$(J) THEN 2ØØ
13Ø LET T$=G$
14Ø LET G$=A$(J)
15Ø LET A$(J)=T$
2ØØ NEXT J
21Ø LET A$(K)=G$
22Ø NEXT K
23Ø PRINT "ALPHABETIZED LIST....."
24Ø FOR B=1 TO N
25Ø PRINT A$(B);" ";
252 IF B/9<>INT(B/9) THEN 26Ø
255 PRINT
26Ø NEXT B
27Ø DATA 17,ADAMS,CONWAY,COHEN,BAILEY,SHEAR,LONG,LEO
31Ø DATA WHALEN,WHEELER,IKE,LONGDO,ROGO,SMITH,JONES
32Ø DATA BENCH,MEEMOM,PEEPOP
4ØØ END
```

ANALYSIS

The program is exactly the same as the numeric sorter except that strings are used instead of real variables.

The computer compares strings by using the coding scheme found in the appendix. For items that do not differ in any letter such as *LONG* and *LONGDO* the computer will print the shorter item first. A blank in a string has a coding number of *32* and is *smaller* than any letter. *A* has a coding number of *65*.

145

With the advent of the touch-tone phone came a new era in musical productions. The tone put out by each button is equivalent to a note from the musical scale. Some phones have twelve buttons, some only ten. A respectable version of many songs can be generated using the proper sequence and timing of tones.

For example a fair version of *Raindrops Keep Fallin' on My Head* can be played by punching out 33363213. *Twinkle, Twinkle, Little Star* will result from tapping out 1199009. Even Beethoven's *Fifth Symphony* can be heard by 0005 8883.

Write a computer program to translate a song from ordinary sheet music into a respectable version suitable for the touch-tone phone. When experimenting be sure to call a friend before you start tapping, or you're liable to wind up with a bill for a long-distance call to East Kinorki.

The notes and their button equivalents are reproduced below.

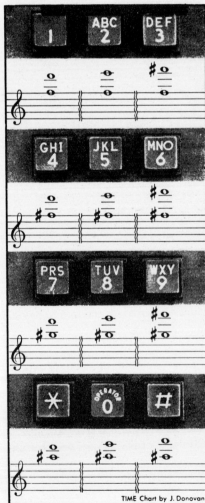

TIME Chart by J. Donovan

There is a trend in music to let the computer compose melodies and even lyrics for certain songs. These programs are rather involved musically speaking and are consistent in any number of aspects.

An interesting experiment would be to allow the computer to generate notes at random on a staff. The only check by the programmer would have to be regarding key and timing.

Have the computer type out a five line staff along with notes imposed upon it. If you wish you can draw the staff beforehand, or even insert a musical sheet sideways. Try to use different symbols for notes of different duration. A clever student might even be able to form the notes as they actually look. [o' or 0 or o' or some other concoction could be made to look like notes.].

Attempts to play the music could lead to interesting interpretations in the right musical mind.

Write a program to produce a facsimile of a check. The face value should be less than $100.

Have the computer accept as INPUT the date, the payee, the signator, the transit numbers and of course the amount of the check. Remember, this value must also appear in alphanumeric form in the space beneath the payee. This may be accomplished by actually typing the amount in words or by translating into alphamerics—a most tedious operation.

Dress your check up with a bank of your own creation. Neatness counts!

The check below is larger than actual size. Use the whole width of the paper.

```
*****************************************************************
*                                                             *
*   EAST KINORKI ROD & GUN CLUB               NO.   69        *
*                                                             *
*                         APRIL 1    1972   29-1              *
*                                            1089             *
*                                                             *
*   PAY TO  EUSTACIUS RIGGLETTS               $   97.34       *
*                                                             *
*   NINETY-SEVEN AND 34/100 ********************* DOLLARS      *
*                                                             *
*   FIFTH NATIONAL BANK                                       *
*      YAMAHOO FALLS                                          *
*                                          X                  *
*   0213-001-13-435-17..                                      *
*****************************************************************
```

Program the computer to play Tic-Tac-Toe.

Allow the player the option of going first or second. Set up a numbered array of the game board and work from there; moves can be referred to by number and the player can do his own bookwork. It takes too much computer time to constantly print out the updated array.

Study the strategy of the game carefully. With proper planning the worst the computer can do is draw. The computers winning record affects the outcome of this problem pointwise.

Allow for input by number and be sure the computer makes it clear exactly how the moves should be entered. Inexperienced people may be running this program.

The computer should be able to declare itself the winner or admit that it's been drawn.

```
 1 | 2 | 3
-----------
 8 | 9 | 4
-----------
 7 | 6 | 5
```

References:
Martin Gardner, *Mathematical Games,* **Scientific American**.

```
THIS PROGRAM PLAYS TIC TAC TOE
THE GAME BOARD IS NUMBERED THUS:
        1    2    3
        8    9    4
        7    6    5


    COMPUTER MOVES  9
    YOUR MOVE?  3
    COMPUTER MOVES  4
    YOUR MOVE?  8
    COMPUTER MOVES  6
    YOUR MOVE?  2
    COMPUTER MOVES  1
    YOUR MOVE?  5
    COMPUTER MOVES  7
    THE GAME IS DRAWN!!


    COMPUTER MOVES  9
    YOUR MOVE?  4
    COMPUTER MOVES  5
    YOUR MOVE?  1
    COMPUTER MOVES  7
    YOUR MOVE?  3
    COMPUTER MOVES  6
    AND WINS *******


    COMPUTER MOVES  9
    YOUR MOVE?  STOP
      PROGRAM STOPPED.
```

```basic
100 PRINT 'THIS PROGRAM PLAYS TIC TAC TOE'
110 PRINT 'THE GAME BOARD IS NUMBERED THUS:'
120 PRINT '1 2 3'
130 PRINT '8 9 4'
140 PRINT '7 6 5'
150 PRINT
180 DEF FNM(X) = X-8*INT((X-1)/8)
200 GO TO 210
210 PRINT
220 PRINT
230 LET A = 9
240 LET M = A
250 GOSUB 650
260 LET P = M
270 LET B = FNM(P + 1)
280 LET M = B
290 GOSUB 650
300 LET Q = M
310 IF Q = FNM(B + 4) THEN 360
320 LET C = FNM(B + 4)
330 LET M = C
340 GOSUB 700
350 GO TO 730
360 LET C = FNM(B + 2)
370 LET M = C
380 GOSUB 650
390 LET R = M
400 IF R = FNM(C + 4) THEN 450
410 LET D = FNM(C + 4)
420 LET M = D
430 GOSUB 700
440 GO TO 730
450 IF P/2 < > INT(P/2) THEN 500
460 LET D = FNM(C + 7)
470 LET M = D
480 GOSUB 700
490 GO TO 730
500 LET D = FNM(C + 3)
510 LET M = D
520 GOSUB 650
530 LET S = M
540 IF S = FNM(D + 4) THEN 590
550 LET E = FNM(D + 4)
560 LET M = E
570 GOSUB 700
580 GO TO 730
590 LET E = FNM(D + 6)
600 LET M = E
610 GOSUB 700
620 PRINT 'THE GAME IS DRAWN!!'
630 GOTO 210
650 GOSUB 700
660 PRINT 'YOUR MOVE';
670 INPUT M
680 RETURN
700 PRINT 'COMPUTER MOVES ';M
710 RETURN
730 PRINT 'AND WINS *******'
740 GO TO 210
750 END
```

A *palindrome* is a word or number which reads the same backwards as forwards.

The word *otto* is a palindrome. So is the phrase, *Able was I 'ere I saw Elba.*

Program the computer to test a phrase to see if it is a palindrome. Input the entire line as a string using a LIN-PUT statement. You can test numbers better than letters so convert the letters to their coded equivalents and test those from end-to-end.

The program should be capable of ignoring punctuation and spacing.

```
? 123454321


THE ABOVE IS A PALINDROME


? 123


THE ABOVE IS NOT A PALINDROME

? 4


THE ABOVE IS A PALINDROME


? OTTO


THE ABOVE IS A PALINDROME


? ABLEWASIEREISAWELBA


THE ABOVE IS A PALINDROME


? PROGRAM


THE ABOVE IS NOT A PALINDROME
```

```
100 REM **** PALINDROME DETECTOR ********
110 DIM A(72),B(72)
120 READ C$
130 DATA ',."-'
140 CHANGE C$ TO C
150 PRINT
160 PRINT
170 LINPUT A$
180 FOR I = 1 TO 2 PRINT
190 CHANGE A$ TO A
200 J = 0
210 FOR I = 1 TO A(0)
220 IF A(I) = C(1) OR A(I) = C(2) OR A(I) = C(3) THEN 260
230 IF A(I) = C(4) OR A(I) = C(5) THEN 260
240 J = J + 1
250 B(J) = A(I)
260 NEXT I
270 K = INT (J/2)
280 FOR I = 1 TO K
290 IF B(I) < > B(J) THEN 340
300 J = J-1
310 NEXT I
320 PRINT 'THE ABOVE IS A PALINDROME'
330 GO TO 150
340 PRINT 'THE ABOVE IS NOT A PALINDROME'
350 GO TO 150
360 END
```

ANALYSIS

The palindromes are entered as strings on line 50. They are converted to their coded numeric equivalents. These are then tested end-to-end for a match.

The loop beginning at 210 tests for the matching ends. In a sense, it folds up the string from the center and checks for symmetry.

Lines 120 through 160 test the string entered for punctuation marks and then tells the system to ignore them by branching to line 190.

81   Metric Recipe Conversion

The United States is the only major industrial nation in the world which does not use the metric system of measurement. The day is not far off when we will convert to such a system. Every quantity will be represented as a decimal part of another. This will considerably simplify computation.

Getting used to such a system will be difficult. In anticipation of that day, write a program to convert a recipe in your cookbook (we certainly can't expect everyone to buy new cookbooks) from the units we use today into the less familiar metric units.

You might also consider writing in the opposite direction; for it is conceivable that outstanding recipes in future cookbooks would have to be bypassed unless you had a way to convert them into units for which you had tools to cook and bake.

A table of the major units in each system is given below.

```
RECIPE CONVERTER                    1 Liter=1.0567 Quart    1 Kilogram = 2.205 Pounds

CONVERTS RECIPES FROM [CUSTOMARY] TO METRIC
USE THESE ABBREVIATIONS:
T=TEASPOON                 TB=TABLESPOON
C=CUP                      OZ=OUNCES
PT=PINT                    QT=QUART
N=NO CONVERSION NECESSARY

WHAT IS THE NAME OF YOUR RECIPE? STUFFED HENS
WHAT IS YOUR NAME? ANDI FRIEDMAN
TYPE QUANTITY, UNIT OF MEASURE,NAME OF ITEM
WHEN DONE, TYPE '0,END,0'
? 6,N,GAME HENS
? .5,C,BUTTER
? 4,C,ONIONS
? 2,C,BREAD CRUMBS
? .25,PT,WINE
? 0,END,0


BAKING TEMPERATURE? 350
BAKE HOW LONG? 45 MINUTES

DIRECTIONS:
? WASH INSIDE OF HENS AND DRAINWELL.  SPRINKLE HENS INSIDE
? AND OUT WITH SALT AND PEPPER.  SPOON STUFFING MIXTURE INTO
? CAVITIES; DO NO PACKING.  SEW OPENING.  ROAST HENS UNCOVERED.
? END

STUFFED HENS                    FROM THE KITCHEN OF ANDI FRIEDMAN


INGREDIENTS:
  6                GAME HENS
  118    ML        BUTTER
  946    ML        ONIONS
  473    ML        BREAD CRUMBS
  118    ML        WINE

WASH INSIDE OF HENS AND DRAINWELL.  SPRINKLE HENS INSIDE
AND OUT WITH SALT AND PEPPER.  SPOON STUFFING MIXTURE INTO
CAVITIES; DO NO PACKING.  SEW OPENING.  ROAST HENS UNCOVERED.
BAKE AT 176.667  DEGREES FOR 45 MINUTES
```

```
10 DIM Q(40),U$(40),I$(40),D$(10)
20 STRINGS 80
30 PRINT
40 PRINT
50 PRINT 'RECIPE CONVERTER'
60 PRINT
70 PRINT 'CONVERTS RECIPES FROM [CUSTOMARY] TO METRIC'
80 PRINT
82 PRINT 'WHAT IS THE NAME OF YOUR RECIPE';
84 INPUT N$
86 PRINT 'WHAT IS YOUR NAME';
88 INPUT K$
90 N = 1
100 INPUT Q(N),U$(N),I$(N)
110 IF U$(N) = 'END' THEN 140
120 N = N + 1
130 GO TO 100
140 N = N-1
150 PRINT
160 PRINT 'BAKING TEMPERATURE';
170 INPUT T
180 LET T = (T-32)*(5/9)
190 PRINT 'BAKE HOW LONG';
200 INPUT T$
210 PRINT
220 PRINT 'DIRECTIONS:'
230 J = 1
240 INPUT D$(J)
250 IF D$(J) = 'END' THEN 280
260 LET J = J + 1
270 GO TO 240
280 LET J = J-1
290 PRINT
300 PRINT
302 PRINT
306 PRINT
310 PRINT N$;TAB(28);'FROM THE KITCHEN OF ';K$
320 PRINT
330 PRINT 'INGREDIENTS:'
335 FOR M = 1 TO N
340 RESTORE
350 FOR L = 1 TO 7
360 READ O$,P$,C
370 IF U$(M) = O$ THEN 420
380 NEXT L
390 PRINT 'ILLEGAL UNIT...DISREGARDED'
400 GO TO 430
420 PRINT INT(Q(M)*C + .5);' ';P$;TAB(8);I$(M)
430 NEXT M
440 PRINT
450 FOR M = 1 TO J
460 PRINT D$(M)
470 NEXT M
480 PRINT 'BAKE AT';T;' DEGREES ';T$
490 PRINT
500 PRINT
510 GO TO 30
520 DATA T,ML,4.929,TB,ML,14.786,C,ML,236.575,OZ,ML,29.572
530 DATA PT,ML,473.15,QT,ML,946.4,N,' ',1
9999 END
```

ANALYSIS

The problem in converting a recipe written in the English system of units to the metric systems involves not only the English-to-metric conversion but conversion of the many units of kitchen-measure to standard units. To keep the program to a reasonable length it is best to restrict the number of English units used, six in the sample program, and provide the conversion factor within the program. The sample program has the conversion factors in the DATA statements, lines 520 and 530.

The design of the program permits the user to enter all quantities in English units, including temperature, and then input the verbal instructions. Time is inputed so that it is available for output but, obviously, does not need conversion.

The ancient chinese game of NIM is a fascinating one. Philosophers of the highest order used to play the game for serious stakes.

Essentially the game goes like this: a pile of stones is placed before each man. The pile may contain any number of stones. The number is known to both players. (Fifteen is a good number to start with).

Each man, at his turn, must take at least one and not more than three stones. Players alternate until the last stone is taken. The player who takes the last stone loses.

Program the computer to play NIM for a pile of fifteen stones, marbles, lacrosse balls or whatever. Extra credit will be given for a program which can play with any number of objects. Make the program interactive, so that someone with little knowledge of the computer or the game can understand it.

References:
Kemeny and Kurtz, **Basic Programming**, pp. 78-81.

```
NEW GAME!

DO YOU WISH TO CHOOSE LIMITS
? NO
RULES FOR THIS GAME ARE*
YOU MAY REMOVE FROM 1 TO
8 PIECES!
WHOEVER REMOVES THE LAST
PIECE LOSES!!!

WE START WITH  65 OBJECTS!

YOUR MOVE? 1
MY MOVE IS  6


NEW TOTAL IS  58


YOUR MOVE? 3
MY MOVE IS  7


NEW TOTAL IS  48


YOUR MOVE? 2
MY MOVE IS  7


NEW TOTAL IS  39


YOUR MOVE? 2
MY MOVE IS  4


NEW TOTAL IS  33


YOUR MOVE? 5
MY MOVE IS  8


NEW TOTAL IS  20


YOUR MOVE? 1
MY MOVE IS  1


NEW TOTAL IS  18


YOUR MOVE? 8
MY MOVE IS  5


NEW TOTAL IS  5


YOUR MOVE? 4
MY MOVE IS  1

*****YOU WIN*****
```

```
05 RANDOMIZE
10 PRINT 'DO YOU WISH TO CHOOSE LIMITS'
20 INPUT A$
30 IF A$ = 'NO' THEN 120
40 PRINT 'THEN INPUT THE MAX CHOICE AND A PILE NUMBER'
50 INPUT K,N
60 GO TO 140
120 LET N = 20 + INT(80*RND(-1))
130 LET K = 5 + INT(10*RND(-2))
140 PRINT 'RULES FOR THIS GAME ARE* '
150 PRINT 'YOU MAY REMOVE FROM 1 TO ';K;' PIECES!'
160 PRINT ' WHOEVER REMOVES THE LAST PIECE LOSES!!!'
170 PRINT
180 PRINT 'WE START WITH ';N;' OBJECTS!'
190 PRINT
210 PRINT
220 PRINT 'YOUR MOVE';
230 INPUT X
240 IF X > K THEN 270
250 IF X < 1 THEN 270
260 IF X < = NTHEN 310
270 PRINT 'ILLEGAL MOVE
280 PRINT
290 GO TO 220
310 LET Z = 0
320 LET N = N-X
330 IF N > 0 THEN 410
340 PRINT '*** I WIN ***'
350 PRINT
360 PRINT
370 PRINT 'NEW GAME!'
380 PRINT
390 GO TO 05
410 LET Z = 0
420 LET Q = INT((N-1)/(K + 1))
430 LET Y = N-1-Q*(K + 1)
440 IF Y = 0 THEN 530
450 LET N = N-Y
460 PRINT 'MY MOVE IS ';Y
470 PRINT
480 IF N = 0 THEN 600
490 PRINT 'NEW TOTAL IS ';N
500 PRINT
510 GO TO 210
530 IF N > 1 THEN 570
550 LET Y = 1
560 GO TO 450
570 LET Y = 1 + INT(K*RND(-3))
580 GO TO 450
600 PRINT '*****YOU WIN*****'
620 GO TO 350
700 END
```

ANALYSIS

This program plays a version of NIM with one pile. The user is allowed the option of selecting the numbers or the computer will generate random rules in lines 120 and 130. The computer selects its move in lines 420 and 430. The reader is invited to verify the strategy.

The computer can be beaten only if it allows the user to go first which it does each time. Notice there are no random numbers generated in the computer phase of the strategy. It can only be beaten by making the proper play on each move. The first move is critical. To win, add the lower and upper limit—for example, 8 + 1 = 9 in the output. Notice that 63 is the closest multiple of 9 less than 65. Add 1 to 63 and make sure the computer takes its turn with one more than a multiple of 9 available each time. Once the first move has been made simply take enough to make the computer's choice and your choice total 9.

The magic numbers for that particular game would be 64, 55, 46, 37, 28, 19, 10. If the computer has that many left on its turn it will lose if you make your total and its total add to 9. Study the sample game in which the user won more carefully.

A safe and educational method for gambling can be devised using the computer. Have the computer generate random numbers. Then devise a technique for translating these numbers into suits and rank. Be sure you keep track of what has been dealt. Perhaps you can teach the computer how to deal for a couple of forms of poker.

Leave as an option the number of cards to be dealt as well as drawn. A sophisticated version of this program could also declare a winner and keep track of the bets and winnings.

Refer to the chart in the Appendix for the relative odds against and values of each type hand. It should be a program that will accomodate up to six people. Be sure the random number generator is set to produce a different set of random numbers each time. Otherwise the deck will never get shuffled.

```
HOW MANY HANDS DO YOU WANT DEALT (MAX 10)
? 2


    HAND #   1


    SEVEN OF HEARTS
    KING OF HEARTS
    SEVEN OF CLUBS
    THREE OF CLUBS
    SEVEN OF SPADES



    HAND #   2


    QUEEN OF HEARTS
    TEN OF HEARTS
    ACE OF HEARTS
    FIVE OF HEARTS
    FIVE OF CLUBS

HOW MANY HANDS DO YOU WANT DEALT (MAX 10)
? 2


    HAND #   1


    EIGHT OF CLUBS
    THREE OF HEARTS
    THREE OF DIAMONDS
    FIVE OF SPADES
    JACK OF CLUBS



    HAND #   2


    SEVEN OF SPADES
    SIX OF DIAMONDS
    TEN OF SPADES
    EIGHT OF SPADES
    KING OF DIAMONDS
```

```
100 DIM X(52),F$(13),S$(10)
110 RANDOMIZE
120 FOR I = 1 TO 51
130 LET X(I) = I
140 NEXT I
150 FOR S = 0 TO 3
160 READ S$(S)
170 NEXT S
180 FOR F = 0 TO 12
190 READ F$(F)
200 NEXT F
210 PRINT 'HOW MANY HANDS DO YOU WANT DEALT (MAX 10)'
220 INPUT N
230 PRINT
240 FOR B = 1 TO N
250 PRINT ' HAND # ';B
260 PRINT
270 FOR A = 1 TO 5
280 LET I = INT(52*RND)
290 LET Y = X(I)
300 IF Y < 0 THEN 280
310 LET X(I) = -1
320 LET S = INT(Y/13)
330 LET F = Y-13*S
340 PRINT F$(F);' OF ';S$(S)
350 NEXT A
360 PRINT
370 PRINT
380 NEXT B
390 DATA DIAMONDS,HEARTS,CLUBS,SPADES
400 DATA TWO,THREE,FOUR,FIVE,SIX,SEVEN,EIGHT,NINE
410 DATA TEN,JACK,QUEEN,KING,ACE
420 END
```

```
HERE IS THE DEAL
  1   JS
  2   3S
  3   5D
  4   6D
  5   8H
DØ YØU WANT ANY CARDS? YES
HØW MANY CARDS? 4
TYPE THE NUMBER ØF THE CARD TØ BE DISCARDED.
? 2
? 3
? 4
? 5
  1                JS
  2                5H
  3                4D
  4                10D
  5                6S
DØ YØU WISH TØ BET,CALL,ØR FØLD? FØLD
YØU NØW HAVE $ 9980
I NØW HAVE $ 10020
HERE IS THE DEAL
  1   9H
  2   6H
  3   JD
  4   QD
  5   8C
DØ YØU WANT ANY CARDS? YES
HØW MANY CARDS? 3
TYPE THE NUMBER ØF THE CARD TØ BE DISCARDED.
? 1
? 2
? 5
  1                AH
  2                2S
  3                JD
  4                QD
  5                2C
DØ YØU WISH TØ BET,CALL,ØR FØLD? BET
I FØLD.
YØU NØW HAVE $ 10000
I NØW HAVE $ 10000
```

```
15 RANDOMIZE
20 R = S = 10000
40 FOR X = 1 TO 9 READ G$(X)
50 DIM C$(52),C(52),S(52)
60 FOR X = 0 TO 51
70 READ C$(X),C(X),S(X)
80 NEXT X
90 FOR X = 1 TO 60
100 A = INT(52*(RND))
110 B = INT(52*(RND))
120 C$(A) = = C$(B)
121 C(A) = = C(B)
122 S(A) = = S(B)
123 NEXT X
124 FOR X = 0 TO 15
125 B = 51-X
126 C$(B) = = C$(X)
127 C(X) = = C(B)
128 S(X) = = S(B)
129 NEXT X
131 E = B = Q = 0
134 R = R-20
135 S = S-20
136 T = T + 40
140 PRINT 'HERE IS THE DEAL'
150 FOR X = 0 TO 4
160 PRINT X + 1;' ';C$(2*X)
163 N$(X) = C$(2*X)
165 N(X,0) = C(2*X)
166 N(X,1) = S(2*X)
171 M(X,1) = S(2*X + 1)
172 M$(X) = C$(2*X + 1)
173 M(X,0) = C(2*X + 1)
180 NEXT X
190 PRINT 'DO YOU WANT ANY CARDS';
200 INPUT A$
210 IF A$ = 'NO' THEN321
220 PRINT 'HOW MANY CARDS';
230INPUT A
235 Z = 10
237 PRINT 'TYPE THE NUMBER OF THE CARD TO BE DISCARDED.'
240 FOR X = 1 TO A
250 INPUT D(X)
270 NEXT X
280 FOR X = 1 TO A
285 D(X) = D(X)-1
290 N(D(X),0) = C(Z)
291 N(D(X),1) = S(Z)
292 N$(D(X)) = C$(Z)
310 Z = Z + 1
320 NEXT X
321 FOR X = 0 TO 4 PRINT X + 1,N$(X)
322 GO SUB 330
323 GO TO 510
330 FOR X = 0 TO 4 P(X) = Q(X) = F(X) = Y(X) = 0
331 TRACE ON
332 FOR X = 0 TO 3
340 FOR Y = X + 1 TO 4
350 IF M(X,0) > M(Y,0) THEN 370
359 M(X,0) = = M(Y,0)
360 M(X,1) = = M(Y,1)
361 M$(X) = = M$(Y)
370 NEXT Y
380 NEXT X
390 FOR X = 0 TO 3 STEP Y(X) + 1
400 FOR Y = X + 1 TO 4
410 IFM(X,0) > M(Y,0) THEN 440
420 Y(X) = Y(X) + 1

440 IF M(X,0) > M(Y,0) + (Y-X) THEN 470
450 P(X) = P(X) + 1
470 M = M
490 NEXT Y
495 F(M(X,1)) = F(M(X,1)) + 1
500 NEXT X
501 F(M(4,1)) = F(M(4,1)) + 1
505 RETURN
510 FOR X = 0 TO 4
530 Q = Y(X) + Q
540 NEXT X
541 IF Q > 0 THEN 680
560 FOR X = 0 TO 3
580 NEXT X
590 FOR X = 0 TO 1
600 IF P(X) = > 3 THEN 750
610 NEXT X
620 FOR X = 1 TO 4
630 M(X,0) = C(Z)
631 M(X,1) = S(Z)
632 M$(X) = C$(Z)
640 Z = Z + 1
650NEXT X
660 GO SUB 330
670 GO TO 1010
680 FOR X = 0 TO 4
690 IF Y(X) > 0 THEN 720
700 M(X,0) = C(Z)
701 M(X,1) = S(Z)
702 M$(X) = C$(Z)
710 Z = Z + 1
720 X = X + Y(X)
721 NEXT X
730 GO SUB 330
740 GO TO 1010
750 IF P(0) = 4 THEN 1010
760 FOR X = 0 TO 3
770 IF P(X) = 4 THEN 1010
780 IF P(X) < 3 THEN 870
790 FOR Y = 0 TO 4
810 M(X,0) = C(Z)
815 M$(X) = C$(Z)
816 Z = Z + 1
820 M(X,1) = S(Z)
840 NEXT Y
850 GO SUB 330
860 GOTO 1010
870 NEXT X
880 IF P(0) = 4 THEN 1010
890 FOR X = 0 TO 1
900 IF P(X) < 3 THEN 1000
910 FOR Y = X + 1 TO 4
920 IF M(X,0) < = M(Y,0) + Y-X THEN 950
930 M(Y,0) = C(Z)
931 M(Y,1) = S(Z)
932 M$(Y) = C$(Z)
950 NEXT Y
960GO SUB 330
970 GO TO 1010
1000 NEXT X
1010 PRINT'DO YOU WISH TO BET,CALL,OR FOLD';
1020 INPUT B$
1021 E = E + 1
1033 IF B$ = 'FOLD' THEN 1040
1034 IF E = 1 THEN 1080
1036 IF B$ = 'BET' THEN 2040
1037 IF B$ = 'CALL' THEN 2240
1040 R = R + T + B
```

```
1050 T = 0
1055 B = 0
1060 GOTO 1360
1080 FOR X = 1 TO 3
1090 Q(Y(X) + 1) = Q(Y(X) + 1) + 1
1091 X = X + Y(X)
1095 NEXT X
1100 IF Q(4) < 1 THEN 1150
1110 B = R
1120 R = 0
1130 D = 20
1135 D$ = 'FOUR OF A KIND'
1140 GOTO 2040
1150 IF Q(3) < 1 THEN 1250
1160 IF Q(2) < 1 THEN 1210
1170 B = R
1180 R = 0
1190 D = 20
1191 D$ = 'FULL HOUSE'
1200 GOTO 2040
1210 B = INT(.5*R)
1220 R = R-B
1230 D = 10
1231 D$ = '3 OF A KIND'
1240 GO TO 2040
1250 IF Q(2) < 2 THEN 1300
1260 B = INT(.25*R)
1270 R = R-B
1271 D$ = '2 PAIR'
1272 D = 10
1280 GOTO 2040
1300 IF Q(2) = 1 THEN 1380
1330 PRINT 'I FOLD.'
1340 S = S + T
1350 T = 0
1360 PRINT'YOU NOW HAVE $';S
1370 PRINT'I NOW HAVE $';R
1375 GO TO 90
1380 IF M(X,0) < 10 THEN 1390
1390 PRINT'WHAT IS YOUR BET';
1391 TRACE ON
1400 INPUT A
1401 PRINT D$
1410 IF A > 100 THEN 1330
1411 D$ = 'PAIR'
1420 M = M
1421 S = S-A
1422 R = R-A
1423 T = T + 2*A
1425 PRINT 'I CALL'
1430 PRINT 'WHAT DO YOU HAVE;'
1440 PRINT'INPUT WHAT YOU HAVE, RANK'
1450 INPUT I$,I
1460 FOR X = 1 TO 9
1470 IF D$ < > G$(X)THEN 1640
1480 IF I$ = G$(X) THEN 1510
1490 PRINT 'I HAVE';D$,W
1495 FOR X = 0 TO 4 PRINT X + 1,M$(X)
1500 GO TO 1040
1510 IF I < W THEN 1490
1520 PRINT 'WHAT IS THE RANK OF YOUR HIGH CARD'
1530 INPUT I
1540 IF I = M(0,0) THEN 1570
1545 IF I > M(0,0)THEN 1660
1550 PRINT 'MY HIGH CARD IS A ';M(0,0)
1560 GO TO 1040
1570 PRINT 'THE HAND IS A TIE '
1610 R = R + B
```

```
1620 B = 0
1630 GO TO 90
1640 IF I$ = G$(X) THEN 1660
1650 NEXT X
1660 FOR X = 0 TO 4
1670 M$(X) = = N$(X)
1680 M(X,0) = = N(X,0)
1690 M(X,1) = = N(X,1)
1710 NEXT X
1720 GO SUB 330
1730 FOR X = 0 TO 3
1740 IF F(X) < 4 THEN 1782
1741 IF P(0) < 4 THEN 1745
1742 X$ = 'STRAIT FLUSH'
1743 GO TO 2000
1745 X$ = 'FLUSH'
1780 NEXT X
1781 GO TO 1810
1782 IF P(0) < 4 THEN 1780
1790 X$ = 'STRAIT'
1800 GO TO 2000
1810 FOR X = 0 TO 4
1820 Q(Y(X) + 1) = Q(Y(X) + 1) + 1
1821 X = X + Y(X)
1830 NEXT X
1840 IF Q(4) < 1 THEN 1870
1850 X$ = 'FOUR OF A KIND'
1860 GO TO 2000
1870 IF Q(3) < 1 THEN 1930
1880 IF Q(2) < 1 THEN 1910
1890 X$ = 'FULL HOUSE'
1900 GO TO 2000
1910 X$ = '3 OF A KIND'
1920 GO TO 2000
1930 IF Q(2) < 2 THEN 1960
1940 X$ = '2 PAIR'
1950 GO TO 2000
1960 IF Q(2) < 1 THEN 1990
1970 X$ = 'PAIR'
1980 GO TO 2000
1990 X$ = 'ZIP'
2000 IF I$ = X$ THEN 1340
2001 PRINT 'YOU ACTUALLY HAD '; TAB(45);X$
2010 PRINT 'YOUTRIED TO CHEAT ME !!'
2020 PRINT 'YOUR MONEY IS FORFEIT'
2030 STOP
2040 FOR X = 4 TO 0 STEP -1
2041 IF Y(X) > 0 THEN W = M(X,0)
2042 NEXT X
2045 PRINT 'WHAT IS YOUR BET';
2050 INPUT A
2060 IF S-A > = 0 THEN 2090
2070 PRINT 'YOU DON''T HAVE THAT MUCH'
2081 IF O + A > = C THEN 2090
2082 PRINT 'YOUR BET IS SHORT OF EVEN A CALL'
2083 GO TO 2045
2090 IF V < = R + B THEN 2093
2091 PRINT 'YOU CAN''T BET OVER WHAT THE HOUSE HAS'
2092 GO TO 2045
2093 O = O + A
2100 T = T + A
2110 V = INT (O-C)
2120 S = INT(S-A)
2123 T = T + V
2125 IF V > 10*D THEN 2280
2140 F = INT(10*(RND))
2150 F = D*(F + 1)
2160 IF F < V THEN 2140
```

```
2170 C = C + F
2180 T = T + F
2185 B = B-F
2190 IF INT(C) = INT(O) THEN 1420
2200 PRINT ' I BET $';F
2210 PRINT 'YOU NOW HAVE $';S
2220 PRINT 'I NOW HAVE $';R + B
2230 GO TO 1010
2240 S = INT(S-C + O)
2250 T = T + INT(C-O)
2260 GO TO 1430
2280 IF V > 16*D THEN 1420
2281 F = INT (6*(RND))
2290 F = D*(F + 1)
2291 F = F + 10*D
2292 GO TO 2281
2299 DATA 'STRAIT FLUSH'
2300 DATA 'FOUR OF A KIND','FULL HOUSE'
2301 DATA 'FLUSH','STRAIT','3 OF A KIND','2 PAIR'
2302 DATA 'PAIR','ZIP'
2400 DATA '2C',2,0,'2D',2,1,'2H',2,2,'2S',2,3
2401 DATA '3C',3,0,'3D',3,1,'3H',3,2,'3S',3,3
2402 DATA '4C',4,0,'4D',4,1,'4H',4,2,'4S',4,3
2403 DATA '5C',5,0,'5D',5,1,'5H',5,2,'5S',5,3
2404 DATA '6C',6,0,'6D',6,1,'6H',6,2,'6S',6,3
2405 DATA '7C',7,0,'7D',7,1,'7H',7,2,'7S',7,3
2406 DATA '8C',8,0,'8D',8,1,'8H',8,2,'8S',8,3
2407 DATA '9C',9,0,'9D',9,1,'9H',9,2,'9S',9,3
2408 DATA '10C',10,0,'10D',10,1,'10H',10,2,'10S',10,3
2409 DATA 'JC',11,0,'JD',11,1,'JH',11,2,'JS',11,3
2410 DATA 'QC',12,0,'QD',12,1,'QH',12,2,'QS',12,3
2411 DATA 'KC',13,0,'KD',13,1,'KH',13,2,'KS',13,3
2412 DATA 'AC',14,0,'AD',14,1,'AH',14,2,'AS',14,3
20000 END
```

ANALYSIS

This program could be more sophisticated. It just deals a five-card hand to as many players as you specify. The name POKER could just as well have been any other card game which can be played with five cards.

Line 110 shuffles the deck. The loop from 120 to 140 keeps track of cards dealt. In 310, if a card has been dealt it is negated. The S loop in 150 establishes the suit and the F loop in 180 the value. These just provide the computer with some *words* or *strings* to type out and translate from the random number generator.

The actual dealing is done in the A loop in 270. Statement 280 generates a random number from 0 to 52. Statement 300 checks to see if the card has already been dealt. Statement 320 can be seen as the one that establishes the suit and line 330 establishes the face value of the card. Line 340 simply prints out the string equivalents of these numbers.

It is obvious that JACK is 11, QUEEN is 12, etc., in our ordinary numbering scheme. Here, however, JACK is the ninth string, QUEEN the tenth, KING the eleventh and ACE the twelfth.

A more sophisticated version which could analyze the hand and declare the winner is also included for the ambitious reader.

84

Visual Computer Timepiece

Write a program that will receive two numbers separated by a comma as input. These numbers will represent the time.

For example, 10,43 could represent 10:43 o'clock.

Have the computer print the twelve numbers of the clock face in the proper order around a circle and have the hands properly positioned so as to indicate the time entered.

Be sure one hand is longer than the other, experiment with different symbols for maximum visibility and effect.

No reference necessary here. Just your own ingenuity. Don't worry about A.M. and P.M. The clock can't tell the difference, so why should you!

Melody Transportation by Computer

Write a program to transpose a melody from one key to any other key.

You'll have to devise a method of entering the original key along with a notation for sharps, flats and octaves.

Notes could be entered as letters which are subscripted. *C1* could be made middle *C* and *C2* the next octave higher, etc.

A more difficult rendition of the problem would involve printing out the transposed music on a staff. That would mean you would have to input the duration of notes and devise a scheme for printing them.

The Magic Century Mark

Take the digits 1 through 9, written in increasing order, and insert either of the following three symbols between them:

+ (addition)   − (subtraction)   blank (run the digits)

Find all the different ways that will produce the arithmetic value of 100.

One example is the one below:

$$1+23-4+56+7+8+9=100$$

Since there are three possibilities and eight spots to be filled, there are 3^8 ways to test, that is 6561 possibilities.

About 10 of them will produce 100.

THE MAGIC CENTURY MARK

```
 1 + 2 + 3 -4 + 5 + 6 + 78 + 9  = 100
 1 + 2 + 34 -5 + 67 -8 + 9  = 100
 1 + 23 -4 + 5 + 6 + 78 -9  = 100
 1 + 23 -4 + 56 + 7 + 8 + 9  = 100
 12 + 3 + 4 + 5 -6 -7 + 89  = 100
 12 + 3 -4 + 5 + 67 + 8 + 9  = 100
 12 -3 -4 + 5 -6 + 7 + 89  = 100
 123 + 4 -5 + 67 -89  = 100
 123 + 45 -67 + 8 -9  = 100
 123 -4 -5 -6 -7 + 8 -9  = 100
 123 -45 -67 + 89  = 100
```

```
95 PRINT "THE MAGIC CENTURY MARK"
96 PRINT
97 PRINT
100 DIM A(9),B(9)
105 FOR I=2 TO 9
110 LET A(I)=0
120 NEXT I
132 FOR I=1 TO 9
134 LET B(I)=I
136 NEXT I
140 FOR I=2 TO 9
150 IF A(I)<>1 THEN 160
152 LET B(I)=-B(I)
154 GOTO 180
160 IF A(I)<>2 THEN 180
162 LET B(I)=SGN(B(I-1))*(ABS(B(I-1))*10+B(I))
170 LET B(I-1)=0
180 NEXT I
190 LET A=0
195 FOR I=1 TO 9
200 LET A=A+B(I)
210 NEXT I
220 IF A<>100 THEN 235
221 LET F=0
222 FOR I=1 TO 9
223 IF B(I)=0 THEN 230
224 IF B(I)<0 THEN 228
225 IF F=0 THEN 228
226 PRINT "+";B(I);
227 GOTO 230
228 PRINT B(I);
229 LET F=1
230 NEXT I
231 PRINT " = 100"
235 LET B(1)=1
236 LET A(9)=A(9)+1
237 FOR I=9 TO 3 STEP -1
250 IF A(I)<3 THEN 265
260 LET A(I)=0
262 LET A(I-1)=A(I-1)+1
263 NEXT I
265 IF A(2)<3 THEN 132
280 END
```

ANALYSIS

The program examines the single digits, 1 to 9, groups (packs) the digits, without reordering, and assigns algebraic signs to spaces not removed by packing. The numbers in the modified pattern are summed algebraically and the sum checked to see if it is 100.

Within the program, $B(I)$ represents the ith number and, $A(I)$ represents the spaces before the ith number. Initially, the subscript vaue of each $B(I)$ is assigned the value of I and all $A(I)$'s are set to zero. An $A(I)$ of 0 represents a +, 1 is -, and 2 is a deleted space. Line 162 increases $B(I)$ to the next factor of 10 when needed, line 200 summing the sequence, and line 220 provides the sum. Output is done by lines 222-230.

$A(I)$ may have one of only three values: 0, 1, or 2, and the value is assigned from the right end of the string. $A(I)$ designates the sign *between* numbers so $A(2)$ is the space between 1 and 2 and, at the other end, $A(9)$ is the space between 8 and 9. The pattern of $A(I)$ is generated as shown below:

0 0 0 0 0 0 0 0 0 yields $1+2+3+4+5+6+7+8+9+0$
0 0 0 0 0 0 0 0 1 yields $1+2+3+4+5+6+7+8-9$
0 0 0 0 0 0 0 0 2 yields $1+2+3+4+5+6+7+89$

and so forth. Note that this is nothing more than a modulo 3 counter. The counting sequence is done in lines 235-265.

165

The neophyte in computer circles is often confused and bewildered by the ASCII code (American Standard Code for Information Interchange). The actual configuration of holes in paper tape is practically undecipherable without a handbook.

Write a program to punch out on paper tape any alphanumeric (worded) expression which you may input. Some BASIC compilers have CHANGE statements which could prove helpful. Some time-sharing companies also have subroutines available to do such things.

It is a difficult and time-consuming task and is not mathematical in its nature. It is however an outstanding exercise in statement manipulation and is not beyond the scope of the talented student.

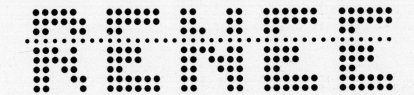

Computer Verse Forms

Give the computer a simple but adequate vocabulary list.

Have the computer compose some short verses of poetry. This can be done by a random selection process or by more careful choosing. Be sure to develop a rhyme scheme or scanning pattern and develop an algorithm to test the grammatical nature of your sentences.

Try to limit your sequences to grammatically possible ones. The Japanese verse form containing exactly 17 syllables, known as *haiku* may be worth investigating.

Your English teacher may wish to collaborate.

One of the Unsolved Problems of Arithmetic

Try to find three integers x, y, and z such that:

$$(x + y + z)^3 = xyz$$

None of the three may equal zero. The problem originally proposed by Werner Mnich, a student at Warsaw University, is to prove that three rationals u, v, and w exist such that:

$$u + v + w = uvw = 1$$

Mnich transformed both of the above equations into an equivalent question—that is, whether there existed integers a, b, and c such that:

$$a/b + b/c + c/a = 1$$

The proof of any of the above would be a solution to one of the unsolved problems of arithmetic.

HINT: Solution of the equation involving integers is to be recommended. Loops are ideally suited to integral problems because it is possible to include all integers between specified limits. Since the rationals and the reals are *everywhere dense* problems of this nature are out.

References:
W. Sierpinski, **Some Unsolved Problems of Arithmetic**.

Two More of the Unsolved Problems of Arithmetic

A couple of problems that look like quickies but really aren't:

1) Do there exist any solutions to the equation
$x^x - y^y = z^z$ where x, y, and z are odd and greater than 1?

2) Do there exist three successive natural numbers each of which is a power of a natural number (exponent greater than 1)?

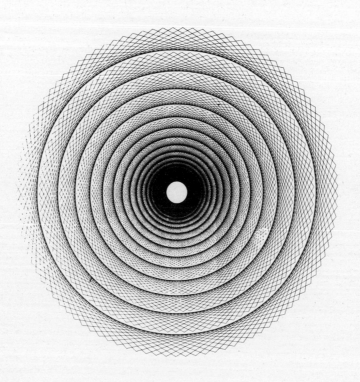

Appendix

A TRIGONOMETRIC GENERATION OF PASCAL'S TRIANGLE

n	sin(nx) + cos(nx)	coefficients
1	$[1] \cos x + [1] \sin x$	1 1
2	$[1] \cos^2 x + [2] \sin x \cos x - [1] \sin^2 x$	1 2 1
3	$[1] \cos^3 x + [3] \sin x \cos^2 x - [3] \sin^2 x \cos x - [1] \sin^3 x$	1 3 3 1
4	$[1] \cos^4 x + [4] \sin x \cos^3 x - [6] \sin^2 x \cos^2 x - [4] \sin^3 x \cos x + [1] \sin^4 x$	1 4 6 4 1
$\vdots$	$\vdots$	$\vdots$

a) This is not simply a remake of the binomial theorem. As it turns out

$$\sin(nx) + \cos(nx) \neq (\sin x + \cos x)^n \text{ for all 'n'.}$$

b) Investigate the following however: the sine and cosine can be related as follows,

$$[\cos(nx) + i \sin(nx)] = (\cos x + i \sin x)^n \text{ where } i^2 = -1.$$

The above is known as DEMOIVRE'S THEOREM!

ASCII Character Code
(Decimal Value)

Decimal Value	ASCII Character		Decimal Value	ASCII Character	
32	SP	SPACE	64	@	
33	!		65	A	
34	"		66	B	
35	#		67	C	
36	$		68	D	
37	%		69	E	
38	&		70	F	
39	'	APOSTROPHE	71	G	
40	(		72	H	
41	)		73	I	
42	*		74	J	
43	+		75	K	
44	,	COMMA	76	L	
45	—		77	M	
46	.		78	N	
47	/		79	O	
48	0		80	P	
49	1		81	Q	
50	2		82	R	
51	3		83	S	
52	4		84	T	
53	5		85	U	
54	6		86	V	
55	7		87	W	
56	8		88	X	
57	9		89	Y	
58	:		90	Z	
59	;		91	[	
60	<		92	\	BACKSLASH
61	=		93	]	
62	>		94	^	or ↑
63	?		95	_	or ←

Summary of Statistical Measures

Measures of Central Tendency:

Arithmetic Mean $= \bar{x} = \dfrac{\sum x_i}{n}$

Median = middle score for n odd sorted scores. Mean of two middle scores for n even.

Mode = the most frequent score.

Mean – Mode $\cong$ 3(Mean – Median)

Geometric Mean $= \sqrt[n]{x_1 \cdot x_2 \cdot x_3 \cdot \ldots \cdot x_n} = G$

Harmonic Mean $= \dfrac{1}{\dfrac{1}{n} \sum \dfrac{1}{x_i}} = H$

Root Mean Square (RMS) $= \sqrt{\dfrac{\sum x_i^2}{n}}$

Note: $H < G < \bar{x}$

Measures of Dispersion:

Range $= x_{max} - x_{min}$ when scores are ordered from highest to lowest.

Mean Deviation from Mean $= \dfrac{\sum |x_i - \bar{x}|}{n}$

Variance $= \sigma^2 = \dfrac{\sum |x_i - \bar{x}|^2}{n}$

Standard Deviation $= \sigma = \sqrt{\text{variance}}$

Coefficient of Variation $= \dfrac{\sigma}{x}$

Mean Deviation from Mean $\cong \dfrac{4\sigma}{5}$

Normal Distribution Dispersion: $68.27\% \equiv \text{mean} \pm \sigma$
$95.45\% \equiv \text{mean} \pm 2\sigma$
$99.73\% \equiv \text{mean} \pm 3\sigma$

Data for Linear Data Regression and Correlation Study

I.Q.	SAT Verbal
132	532
126	538
127	591
117	446
127	538
120	433
125	696
125	591
133	506
119	499
122	519
134	650
123	525
125	387
122	519
120	446
132	519
134	492
126	637
141	598
141	545
114	420
110	486
141	486
128	611
118	453
123	453
128	644
121	571
121	578
122	433
125	578
126	552
132	630
132	552
134	708
127	506
121	512
138	650
125	499
132	519
108	511
121	446
128	677
123	617
126	644
125	584
122	479
120	519
133	571

$n = 50$

Note: It is important that the scores remain paired for the correlation study. It is necessary to order the scores for a regression analysis involving median and mode, but you can only order one set of scores for a correlation.

Close inspection of the formula for correlation will lead you to conclude that order is unimportant unless the analysis is done by *ranking*.

Odds Against Drawing A Certain Poker Hand

HAND	NUMBER POSSIBLE	ODDS
Royal Flush	4	649,739 to 1
Straight Flush	36	74,192 to 1
Four of a Kind	624	4,164 to 1
Full House	3,744	693 to 1
Flush	5,108	508 to 1
Straight	10,200	254 to 1
Three of a Kind	54,912	46 to 1
Two Pair	123,552	20 to 1
One Pair	1,098,240	4 to 3
Nothing	1,302,540	Even
All Possible Hands	2,598,960	

176

ASCII CODE FOR EIGHT CHANNEL PAPER TAPE

(EVEN PARITY)

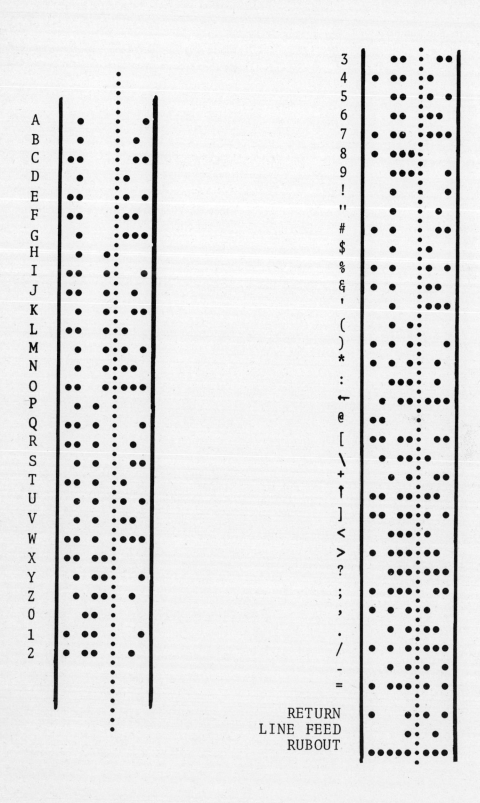

Programming Tricks

1. To sum a group of numbers:

   `10  S=S+A`

   where **A** is the current value of the variable and **S** is the partial sum.

2. Test for divisibility:

   `20  IF X/K=INT(X/K) THEN ....`

   is true when **X** is divisible by **K**.

3. To compute the values after division:

   `30  R=X-INT(X/A)`

   makes **R** the remainder when **X** is divided by **A**.

4. To exchange the values of two variables:

   ```
   40  S=10
   50  T=12
   60  W=S
   70  S=T
   80  T=W
   ```

   Lines 60, 70 and 80 execute the exchange by assigning the value of **S** to **W**, **T** as the new **S**, and **W** (old **S**) as **T**.

5. To compute small factorials:

   ```
    90  I=1
   100  FOR X=1 TO N
   110  I=I*1
   120  NEXT X
   130  END
   ```

   Factorial **N** will be the final value of **I**.

6. To exit from a program with a string test:

   ```
   140  INPUT A$
   150  IF A$="YES" THEN STOP
   ```

   Execution will stop when "Yes" is input.

7. To decompose an integer into its digits:

   ```
   160  X=123
   170  A=INT(X/100)
   180  B=INT(./*(X-100*A)
   190  C=X-100*A-10*B
   ```

   A is hundreds digits of **X**,
   B is tens digits of **X**,
   C is units of **X**.

8. To compute compound interest directly:

   ```
   200  FOR X=1 TO N
   210  I=P*R
   220  P=P+I
   230  NEXT X
   ```

   where **R** is rate of interest,
   I is amount of interest and
   P is the principal.

9. To generate random numbers within a certain range:

   ```
   240  RANDOMIZE
   250  T=INT(((B—A)+I)*RND)+A
   ```

   B must be greater than **A**.

10. To round numbers:

    ```
    260  X=(INT(X+.5)
    270  X=(INT(10*X+.5))/10
    ```

    X to nearest integer.
    X to nearest tenth.

11. To convert angle measure:

     290   R=D/57.295779   **D** degrees to **R** radians.
     300   D=57.295779*R   **R** radians to **D** degrees.

12. To compare two numbers when truncation errors are likely:

     310   IF ABS(B—X)<.00001 THEN ...   will act as though **B** equals **X**.

13. To conserve paper when printing:

     320   PRINT B(X);   will print multiple values of **B(X)** on a line.

14. To find the area of a triangle when only its sides are known:

     330   S=(A+B+C)/2   where **A**, **B**, and **C** are sides, **T** is area.
     340   T=SQR(S*(S—A)*(S—B)*(S—C))

15. To find the antilogs of a base 10 logarithm:

     350   Y=10↑X   **Y** is antilog of **X**.

16. To convert the base of logarithms:

     360   DEF FNL(X)=LOG(X)/LOG(10)   gives base 10 logarithms when LOG is base *e*.
     370   DEF FNL(Y)=LOG(Y)/LOG(A)   gives base **A** logarithms when LOG is logarithm to any other base **Y**.

17. When multiplying fractions do not multiply numerators and denominators and then divide: divide first and then multiply the decimals.

18. Do not try to decompose a number into its individual digits when it is possible to recompose the digits back to the number.

Bibliography

Albrecht, R.L., Finkel, L. and Brown, J.R. **Basic For Home Computers**. New York: John Wiley and Sons, 1978. ($6.95)

Ahl, David H. **Basic Computer Games**. Morristown, N.J.: Creative Computing Press, 1978. ($7.50)

_____. **The Best of Creative Computing, Volume 1**. Morristown, N.J.: Creative Computing Press, 1976. ($8.95)

_____. **The Best of Creative Computing, Volume 2**. Morristown, N.J.: Creative Computing Press, 1977. ($8.95)

_____. **More Basic Computer Games**. Morristown, N.J.: Creative Computing Press, 1979. ($7.50)

Alder, Henry. **Introduction to Probability and Statistics**. San Francisco: W.H. Freeman and Co., 1962.

Amir-Moez, A.R. "Rational Approximations to π." **More Chips from the Mathematical Log**. Norman, Okla.: University of Oklahoma Press, 1970.

Asimov, Isaac. **Understanding Physics: Motion, Sound and Heat**. New York: New American Library, 1969.

Bakst, Aaron. **Mathematical Puzzles and Pastimes**. Princeton: D. Van Nostrand Co., Inc., 1965.

Ball, W. **Mathematical Recreations and Essays**. London: MacMillan and Co., 1914.

Bashaw, W.L. **Mathematics for Statistics**. New York: John Wiley and Sons, 1969.

Beiler, A.H. **Recreations in the Theory of Numbers**. New York: Dover Publishing Co., 1970.

Caldwell, J.H. **Topics in Recreational Mathematics**. Cambridge: Cambridge University Press, 1966.

Calingaert, Peter. **Principles of Computation**. Reading, Mass.: Addison-Wesley Publishing Co., 1965.

Davis, Phillip. **Lore of Large Numbers**. New York: Random House, 1961.

Dorrie, Heinrich. **100 Great Problems of Elementary Mathematics**. New York: Dover Publishing Co., 1965.

Dwyer, Thomas and Critchfield, Margot. **Basic and the Personal Computer**. Reading, Mass.: Addison-Wesley Publishing Co., 1978. ($12.95)

_____. and Kaufman, Michael S. **A Guided Tour of Computer Programming in Basic**. Boston: Houghton Mifflin Company, 1973. ($6.15)

Ehle, B.R. **Elementary Computer Applications**. New York: John Wiley and Sons, 1971.

Gardner, Martin. "Mathematical Games." **Scientific American**. Vol. 225, No. 2, August 1971.

Greenspan, Donald. **Introduction to Calculus**. New York: Harper and Row, 1968.

Gruenberger, Fred and Jaffray, George. **Problems for Computer Solution**. New York: John Wiley and Sons, 1965. ($13.50)

Keedy, M. **Algebra and Trigonometry**. New York: Holt, Rinehart and Winston, 1967.

Kemeny, John G. and Kurtz, Thomas E. **Basic Programming**, Second Edition. New York: John Wiley and Sons, 1965. ($11.95)

Khinchin, A. Y. **Continued Fractions**. Chicago: University of Chicago Press, 1964.

Kirch, A. M. **Elementary Number Theory: A Computer Approach**. New York: Intext, Inc., 1974.

Kraitchik. **Mathematical Recreations**. New York: W.W. Norton and Co., 1942.

Lee, Elvin J. "History and Discovery of Amicable Numbers." **Journal of Recreational Mathematics**. April, 1972.

Lee, John. **Numerical Analysis for Computers**. New York: Reinhold Publishing Co., 1966.

LeVeque, William. **Elementary Theory of Numbers**. Reading, Mass.: Addison-Wesley, 1962.

Loewen, Kenneth, "Matrices." **More Chips from the Mathematical Log**. Norman, Okla.: University of Oklahoma Press, 1970.

Mancill, Julian. **Modern Analytical Trigonometry**. New York: Dodd, Mead and Company, 1960.

Meyer, J.S. **Fun with Mathematics**. Cleveland: World Publishing Co., 1952.

————. **More Fun with Mathematics**. Cleveland: World Publishing Co., 1952.

Moise, Edwin. **Geometry**. Reading, Mass.: Addison-Wesley Publishing Co., 1967.

Mostellier, Fred. **Probability and Statistics**. Reading, Mass.: Addison-Wesley Publishing Co., 1961.

Newman, James. **The World of Mathematics**, 4 volumes. New York: Simon and Schuster, 1956.

Nivan, Ivan. **Introduction to Computing Through the Basic Language**. New York: Holt, Rinehart and Winston, 1969.

Ore, O. **Invitation to Number Theory**. New York: Random House, 1967.

Rogowski, Stephen J. **Computer Clippings**. Palo Alto, CA: Creative Publications, 1975.

Ruckstahl, Kathy and Wilford, Charles. "Prime Numbers: Yes, No, Perhaps." **More Chips from the Mathematical Log**. Norman, Okla.: University of Oklahoma Press, 1970.

Sage, Edwin R. **Fun and Games with the Computer**. Newburyport, Mass.: Entelek, Inc., 1975. ($8.95)

————. **Problem-Solving with the Computer**. Newburyport, Mass.: Entelek, Inc., 1969. ($8.95)

Sears and Zemansky. **University Physics**. Reading, Mass.: Addison-Wesley Publishing Co., 1963.

Selby, Samuel. **CRC Standard Mathematical Tables**. Cleveland: Chemical Rubber Co., 1965.

Sierpinski, W. "Some Unsolved Problems of Arithmetic." **28th Yearbook of the NTCM**. Washington, D.C., 1963.

Spencer, Donald D. **Fun with Computers and Basic**. Ormond Beach, Fla.: Camelot Publishing Co., 1978. ($6.95)

————. **Using Basic in the Classroom**. Ormond Beach, Fla.: Camelot Publishing Co., 1978. ($9.95)

Stein, Edwin. **Fundamentals of Mathematics**. Boston: Allyn and Bacon, 1967.

Stewart, B.M. **Theory of Numbers**. New York: MacMillan and Co., 1964.

Taylor, Jason. **The Calculus with Analtytical Geometry Handbook**. Bedford, Mass.: Taylor Associates, 1976. ($2.95)

Townsend, Charles Barry. **Merlin's Puzzlers**. Maplewood, N.J.: Hammond, Inc., 1976. (Two volume set $7.50)

Wickelgren, Wayne A. **How to Solve Problems**. San Francisco: W.J. Freeman and Company, 1974. ($7.50)

Wilson, W.A. **Analytic Geometry**. Boston: D.C. Heath and Co., 1949.

Wisner, Robert. **A Panorama of Numbers**. Glenview, Ill.: Scott Foresman and Co., 1970.

Creative Computing can help you select the best computer and get the most out of it.

With so many new personal computers being announced and the prices coming down so rapidly, isn't the best bet to wait a year or so to buy a system?

We think not. A pundit once observed that there are three kinds of people in the world: 1) those who make things happen, 2) those who watch things happen and 3) those who wonder what happened. Today, it is those who are getting involved with microcomputers who are making things happen by learning to use computers effectively.

Furthermore, it is not likely that we will see the same dramatic price declines in future years that have already taken place. Rather, one will be able to get more capability for the same price.

The TI-99/4 has excellent color graphics and costs $1150 including color TV monitor.

Which system is for you?

No two people have exactly the same needs. You'll have to determine what capabilities are important to you. Key variables include:

• Upper and lower case. Obviously vital if you are planning to do word processing or anything with text output.

• Graphics. Most systems have graphics but the resolution varies widely. How much do you really need?

• Color. Some systems are B&W, some have 4 colors, others up to 256 colors. Many colors sounds nice, but do you really need 4, or 16, or more?

• Mass storage. The smaller systems are cassette based; larger systems offer floppy disks or even hard disks. What size data bases do you intend to use and is it important to have high-speed random access to an entire data base?

• Languages. Basic is standard but increasingly Pascal, Fortran, Cobol and special purpose languages are being offered.

• Audio, Speech, Music. Are these features important for your planned applications?

• Applications Software. Third party software is widely available for some systems, non-existent for others. Do you need this, or can you write your own?

Unbiased, in-depth evaluations.

At Creative Computing, we obtain new systems as soon as they are announced. We put them through their paces in our Software Center and also in the environment for which they are intended — home, business, or school. We published the first in-depth evaluations of the Texas Instruments 99/4, Atari 800, TRS-80, Ohio Scientific Challenger, Exidy Sorcerer, Apple II disk system and Heath H-8. We intend to continue this type of coverage, not only of systems, but peripherals and software as well.

Sorting: A Key Technique

While evaluations are important, the main focus of Creative Computing magazine is computer applications of all kinds. Many of these require that data be retrieved or sorted. Unfortunately, most programming texts focus on the bubble sort (or straight insertion) and, very infrequently, another technique (usually delayed replacement) and let it go at that.

Yet, except for comparison counting, the bubble sort is the least efficient. Tutorials and articles in Creative Computing demonstrate that the Shell-Metzner and Heapsort are from 50 to 13,000 times as fast as the bubble sort! Consider a sort of 100,000 items on a DEC System 10:

Bubble sort	7.1 days
Delayed replacement	3.8 days
Heapsort	17.3 minutes
Shell-Metzner	15.0 minutes

Needless to say, on a microcomputer, a bubble sort of even 1000 items is agonizingly long.

Free Sorting and Shuffling Reprint

Because sorting and shuffling (mixing a list of items) is so vital in most programming, we are making available a 20-page reprint booklet on Sorting, Shuffling and File Structures along with our May 1979 issue which has several articles on writing user-oriented programs and making the most of available memory space. The reprint booklet and issue are free with 12-issue or longer subscriptions.

At Creative Computing, we believe that computers can be of benefit to virtually every intelligent person in the

Free reprint booklet and issue with a new subscription to Creative Computing.

Contributing editor Ted Nelson (L) is author of "Computer Lib/Dream Machines." Publisher David Ahl (R) is a pioneer in computer models, simulations and games.

country. We do not believe that the "Computer priesthood" should confuse and bully the public. As Ted Nelson stated in the Computer Lib Pledge, we do not treat any question as a dumb question, since there is no such thing. We are against computer terms or systems that are oppressive, insulting or unkind, and we are doing the best we can to improve or replace such terminology or systems. We are committed to doing all we can to further human understanding and make computers easy to understand, interactive wherever possible, and fun for the user. The complete Computer Lib Pledge is contained in our May 1979 issue which we are furnishing free to new subscribers.

Computer literacy to everyone

The Creative Computing Software Division is participating with Children's Television Workshop in an important new venture, Sesame Place. These theme parks are being designed to bring interactive computer games and simulations to young children (and their parents) and remove the mystique of computers from the youngest segment of our population. In addition, we are participating in projects with several school systems and museums to write reading comprehension and ecology simulations software. We are also involved in a major college-level computer literacy project.

As a subscriber to Creative Computing, you will benefit from all of these activities. Creative Computing is the Number 1 software and applications magazine. Subscribe today — 12 issues for $15 ($9 saving over the newsstand price). Or, beat inflation and get 36 issues for just $40. Money back if you're not satisfied. Send payment or Visa, Master Charge or American Express number to:

Creative Computing
P.O. Box 789-M
Morristown, NJ 07960
Save time, and call your order toll-free to:
800-631-8112
(In NJ call 201-540-0445)

creative computing

The BEST of Creative Computing VOLUME 1

Partial Listing of Contents -

In this 328 page book are all the articles, stories, learning activities, games and puzzles that appeared in Creative Computing Volume 1. The contents cover the gamut of computer applications in education and recreation. Over 200 contributors are represented from college professor to high school student, from U.S. Senator to underground cartoonist and from corporation president to science fiction author. A must for anyone concerned with the role of and potential for the computer in society.

The contents are so diverse and numerous there's room here for only a sampling.

Edited by David Ahl. Large format paperbound, 328 pages, $8.95. (6A)

The BEST of Creative Computing VOLUME 2

Partial Listing of Contents -

336 pages of the best articles, fiction, foolishness, puzzles, programs, games and reviews from Creative Computing Volume 2. A diversity of information and activities so staggering it may well be the "one book must" you need for your reference and to recommend to your friends. A potpourri of information on languages and programming theory, on artificial intelligence, on computers in education and in the arts. 67 pages are devoted to puzzles, programs and things to do. The reviews alone could make the book.

A sampling of the diverse contents is listed.

Edited by David Ahl. Large format, 336 pages, $8.95. (6B)

Katie and the Computer

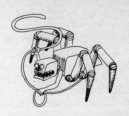

Fred D'Ignazio and Stan Gilliam have created a delightful picture book adventure that explains how a computer works to a child. Katie "falls" into the imaginary land of Cybernia inside her Daddy's home computer. Her journey parallels the path of a simple command through the stages of processing in a computer, thus explaining the fundamentals of computer operation to 4 to 10 year olds. Supplemental explanatory information on compu-

ters, bytes, hardware and software is contained in the front and back end papers.

Thrill with your children as they join the Flower Bytes on a bobsled race to the CPU. Share Katie's excitement as she encounters the multi-legged and mean Bug who lassoes her plane and spins her into a terrifying loop. Laugh at the madcap race she takes with the Flower Painters by bus to the CRT.

Written by Fred D'Ignazio and illustrated in full color by Stan Gilliam. 42 pages, casebound, $6.95. (12A)

A t-shirt with the Program Bug is available in a deep purple design on a beige shirt. Adult size S,M,L, XL. Children's size S,M,L. $5.00.

An Introduction for CHILDREN and beginners to the World of Computers

Be a Computer Literate

Used as a text in many schools, this informative, full color book is an ideal first introduction to the world of computers for children aged 10 to 16. The book is divided into eight chapters:

1 Introduction
2 What are computers
3 Kinds of computers
4 What goes on inside computers
5 Communicating with the computer
6 Language of the computer
7 How to write a simple program
8 How computers work for us.

The full color drawings, diagrams and photos found on every page of these chapters, coupled with the large type, make the book easy to read and understand.

The book contains brief explanations of how computers are used in over sixteen different fields, from medicine to law enforcement, art to business. transportation to education.

The simple glossary provided will help familiarize beginners with essential computer terminology.

Written by Marion J. Ball and Sylvia Charp. Large format, paperbound, 66 pages, $3.95 (6H)

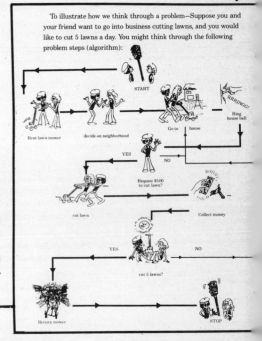

To illustrate how we think through a problem—Suppose you and your friend want to go into business cutting lawns, and you would like to cut 5 lawns a day. You might think through the following problem steps (algorithm):

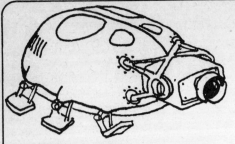

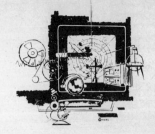

Artist and Computer

This unique book by Ruth Leavitt covers the latest techniques in computer art, animation and sculpture. In its pages 35 artists explain how they use computers as a new means of self-expression. **The San Francisco Review of Books** said "Get yourself a copy of this book if you enjoy feeding your mind a diet of tantalizing high-impact information." Over 160 illustrations, some in full color. 121 pages hardbound [6E] $10.00. Softbound [6D] $4.95.

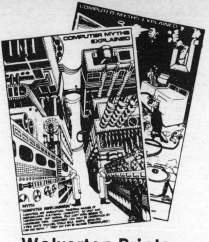

Wolverton Prints

Set of 8 computer Myths Explained by Monte Wolverton. On heavy stock, large 12X17" size, suitable for framing, dressing up that drab line printer or file cabinet. Only $3.00 [5G]

The Colossal Computer Cartoon Book

The best collection of computer cartoons ever! 15 chapters of several hundred cartoons about robots, computer dating, computers in the office, etc. Great gift item. 128 pp. softbound $4.95 [6G]

ORDER FORM

creative computing

P.O. Box 789-M   Morristown, NJ 07960

Please use this order form for fast, dependable service. It gives us the information necessary to insure prompt delivery.

To make payment: We gladly accept your personal check, bank draft, money order, VISA, Master Charge or American Express.
Please do not send currency. Sorry, no C.O.D. orders.

Be sure to include the complete number and expiration date of your card. Your purchase will be included on your regular monthly statement.

Name_____

Address_____ Apt. #_____

City_____ State_____ Zip_____

Area code_____ Telephone_____

Ship to: (if other than yourself)

Name_____

Address_____ Apt. #_____

City_____ State_____ Zip_____

☐ Check or money order enclosed (U.S. funds only)
☐ VISA   ☐ Master Charge   ☐ American Express

_____
Card number

_____   _____
Expiration Date   Signature

Order Toll Free in continental U.S.
800-631-8112
(In NJ call 201-540-0445)

Payment for telephone orders must be made with Visa, MasterCharge, or American Express.

SUBSCRIPTIONS

	USA	Canada and Foreign Surface	Foreign Air
3-year(36 issues)	☐ $40	☐ $67	☐ $130
2-year(24 issues)	☐ $28	☐ $46	☐ $88
1-year(12 issues)	☐ $15	☐ $23	☐ $45

Quantity	Cat. No.	Title	Price	Total
_____	12D	Computers in Mathematics: A Sourcebook of Ideas	$15.95	_____
_____	10R	Computer Coin Games	$3.95	_____
_____	9Y	Problems for Computer Solution	$9.95	_____
_____	12E	The Impact of Computers on Society and Ethics: A Bibliography	$17.95	_____
_____	6H	Be A Computer Literate	$3.95	_____
_____	6A	The Best of Creative Computing: Volume 1	$8.95	_____
_____	6B	The Best of Creative Computing: Volume 2	$8.95	_____
_____	6C	Basic Computer Games	$7.50	_____
_____	6C2	More Basic Computer Games	$7.50	_____
_____	6F	The Best of Byte	$11.95	_____
_____	12A	Katie and the Computer	$6.95	_____
_____	CR101	Computer Music Record	$6.00	_____
_____	6D	Artist and Computer	$4.95	_____
_____	6E	Artist and Computer (hardbound)	$10.00	_____
_____	5G	Wolverton Prints	$3.00	_____
_____	6G	The Colossal Computer Cartoon Book	$4.95	_____

Shipping and handling $2.00

Prices subject to change without notice.

NJ residents add 5% tax   _____

TOTAL   _____